Integrated weed management strategies to

control herbicide resistant

Alopecurus myosuroides Huds.

Integrated weed management strategies to control herbicide resistant *Alopecurus myosuroides* Huds.

Dissertation

in fulfilment of the requirements for the degree

"Doktor der Agrarwissenschaften"

(Dr. sc. Agr.)

to the Faculty of Agricultural Sciences

University of Hohenheim

Institute of Phytomedicine (360)

Department of Weed Science (360b)

Prof. Dr. Roland Gerhards

by

Alexander Kurt Zeller

Born in: Freiburg im Breisgau, Germany

2019

Bibliographical information held by the German National Library

The German National Library has listed this book in the Deutsche Nationalbibliografie (German national bibliography); detailed bibliographic information is available online at http://dnb.d-nb.de.

1st edition - Göttingen: Cuvillier, 2020

Zugl.: D100, Hohenheim, Univ., Diss., 2020

Nonnenstieg 8, 37075 Göttingen, Germany

Telephone: +49 (0)551-54724-0

Telefax: +49 (0)551-54724-21

www.cuvillier.de

1st edition, 2020

This publication is printed on acid-free paper.

ISBN 978-3-7369-7278-0

eISBN 978-3-7369-6278-1

Examination Committee

Supervisor and Reviewer:	Prof. Dr. Roland Gerhards, University of Hohenheim
Co-Reviewer:	Prof. Dr. Jan Petersen, TH Bingen
Additional Examiner:	Prof. Dr. Wilhelm Claupein, University of Hohenheim
	PD. Dr. Regina Belz, University of Hohenheim
Date of oral examination:	17.07.2020

I. Table of Contents

II. List of Figures

III. List of Tables

IV. List of Equation

V. List of Abbreviations

ACCase	Acetyl-CoA carboxylase.
ACE	*Alopecurus myosuroides* control efficacy.
AG	Stock company.
AI	Active ingredient.
ALS	Acetolactate synthase.
ANOVA	Analysis of variances.
ASL	Above sea level.
Asn	Asparagine.
α	Significance level.
BBCH	Bayer, BASF, Ciba-Geigy and Hoechst – Scale, for growth stage identification.
Benth.	George Bentham (* 22. September 1800 in Stoke, Devon England; † 10. September 1884 in London).
CR	Crop rotation.
cv.	Crop variety.
Cys	Cysteine.
°C	Degree Celsius.
E	East.
et al.	And others (Latin, et alii, aliae, alia).
EU	European Union.
€	Euro.

FAO	Food and Agriculture Organization of the United Nations.
g	Gram.
GmbH	Limited liability company.
ha	Hectare.
His	Histamine.
HRAC	Herbicide Resistance Action Committee.
HS	Herbicide strategy.
Huds.	William Hudson (* 1730 in Kendal; † 23. Mai 1793).
Ile	Isoleucine.
IWM	Integrated weed management.
kPa	Kilo pascal.
KTBL	Board of Trustees for Technology and Construction in Agriculture.
L	Liter.
L.	Carl von Linné (* 23. Mai 1707 in Råshult, Älmhult; † 10. Januar 1778 in Uppsala).
Leu	Leucine.
m	Meter.
mm	Millimeter.
MOA	Mode of action.
N	North.
NTSR	Non-target-site resistance.

OECD	Organization for Economic Cooperation and Development.
p	Probability -value.
Pro	Proline.
PSII	Photosystem – II.
SD	Standard Deviation.
t	Ton.
T	Tillage measure.
Trp	Tryptophan.
TSR	Target-site resistance.
Val	Valine.
%	Percent.

Vivat, crescat, floreat L. Württembergia in aeternum.

Chapter I

–

General Introduction

1 General Introduction

According to the latest UN estimates, the human population is projected to grow to 9.7 billion people in 2050 and 10.9 in 2100. This corresponds to an increase of 26 % and 42 % compared to 2019, respectively (United Nations, 2019). Alexandratos and Bruinsma (2012) calculated a 60 % higher demand of agricultural products in 2050 as indicated in the interim report "World Agriculture: towards 2030/2050" of the FAO (FAO, 2006). It is therefore essential to enhance agricultural production. Due to modern machinery, competitive and high-yielding cultivars, fertilizers, irrigation systems and pesticides, the agricultural industry attempts to fulfill the growing demand. However, the progressive sealing of land is opposed to the increasing demand and abiotic factors can cause high yield losses. Additionally, the biotic factors weeds, pests and diseases permanently endanger crop yield (Oerke and Dehne, 2004). In a global point of view, weeds generate the highest biotic yield losses of 34 % by competing with the crop for relevant growth parameters such as space, water, nutrients and light (Oerke, 2006). The time of occurrence and the infestation level of weeds determine the potential damage considerably. Further, weeds also impede harvest, can reduce the product quality and might serve as hosts for pests and diseases (Zwerger and Ammon, 2002).

The competition by weeds can be reduced or even eliminated due to integrated weed management (IWM) strategies, using preventive, cultural, biological, mechanical and chemical weed control measures. Preventive and cultural measures aim to avoid and suppress the occurrence of weeds in a field and support the competitiveness of the crop towards weeds (Swanton and Weise, 1991; Mortensen et al., 1995; Buhler, 2002). Traditionally wide crop rotations were used to maintain soil fertility and to control weeds and pests. However, as a result of the development of pesticides and synthetic fertilizers, most crop rotations have been simplified and narrowed (Leighty, 1938; Froud-Williams, 1988). The mechanical weed control focuses on physical measures, disrupting weed germination and destroying plant tissue. Typical mechanical measures include the cultivation of the crop as well as pre- and post- cultivating tillage but also mowing, cutting and hand weeding (Swanton and Weise, 1991; Buhler, 2002; Rueda-Ayala et al., 2010). Nevertheless, chemical treatments achieve the highest weed control efficacies. Due to the application of selective herbicides, control efficacies of up to 99.99 % can be achieved (Foster et al., 1993). Further, herbicides are easy to use, relatively cheap and have a high area output. From a global perspective, the use of herbicides is the most common and efficient weed control strategy, which has replaced almost all other weed control strategies in conventional cropping systems since they were developed (Heap, 2014). Alternative IWM strategies have been less pursued due to the high control efficacy of herbicides and as a consequence weed pressure and infestations increased unnoticeably.

Through intensive and continuous use of herbicides, a high selection pressure was exerted over a long period. This resulted in a selection of herbicide resistant weed biotypes (Powles and Yu, 2010). Ryan (1970) reported the first well-documented case of herbicide resistance of a triazine-resistant common groundsel *Senecio vulgaris* L. (Asterales: *Asteraceae*) in 1968. To date, about 50 years later, 502 cases in 258 species (150 dicotyledonous and 108 monocotyledonous) were documented worldwide, which prove a propagation of herbicide resistance (Heap, 2019). Herbicide resistance is defined as a naturally occurring and inherent ability of a weed biotype within a weed population, to survive a herbicide application at a rate normally lethal to the wild type and reproduce itself (Powles and Preston, 1995; Heap, 2005). There are different mechanisms known in plants leading to herbicide resistance, that can be divided into target-site (TSR) and non-target-site resistance (NTSR). TSR mechanisms comprise structural changes of target enzymes which stop the herbicide molecule from efficiently binding to the site of action. Further TSR mechanisms are an overexpression of the target site, when the plants synthesize the target protein in larger amounts, or if structural changes in the promotor region of the gene occur (Devine and Shukla, 2000; Gaines et al., 2010). NTSR mechanisms, include the ability of weeds to reduce the herbicide concentration reaching the target enzyme, for example by enhanced metabolism or sequestration of the herbicide molecules into the cell wall or vacuole (Délye, 2013; Délye et al., 2013; Yu and Powles, 2014). Theoretically, resistances evolve towards one active ingredient (AI), however as a result of cross-pollination resistances to different chemical families within one mode of action (MOA) and different MOA can accumulate in one biotype. These types of resistances are considered as cross – and multiple – resistances, respectively (Powles and Preston, 1995; Werck-Reichhart et al., 2000; De Prado and Franco, 2004). The term multiple resistance is also used when more than one mechanism of resistance is present in one biotype (Jutsum and Graham, 1995).

Blackgrass, *Alopecurus myosuroides* Huds. (Poales: *Poaceae*), is a grass weed mainly germinating in autumn and promoted by growing winter annual crops (Naylor, 1972a; Moss, 1985). It is favored by crop rotations with high proportions of winter annual crops, early sowing dates of winter cereals and reduced tillage practices (Moss, 1985; 1987a; Hurle, 1993; Melander, 1995; Gerhards et al., 2013; Lutman et al., 2013). Due to non-persistent seeds, high growth rates and cross-pollination *A. myosuroides* is a prone species for a fast resistance development (Mortimer et al., 1992; De Prado and Franco, 2004). Cross- and multiple-resistant populations are reported throughout Europe and *A. myosuroides* is stated to be one of the most important weed species (Moss et al., 2007; Heap, 2019). In addition to the increasing resistance development, a further challenge is that the number of AIs that are available to control *A. myosuroides* is declining because several AIs lose their permission in the course of the re-registration process (Drobny, 2016).

1.1 Objectives

The aim of this study was to investigate reliable management methods to control *A. myosuroides* effectively. Different IWM strategies were tested. Thereby preventive, cultural, mechanical and chemical weed control methods were conducted and combined. Effect on crop yield, *A. myosuroides* abundance, herbicide control efficacies and the resistance development were investigated. Further, the costs of the IWM strategies were analyzed and the impact of the contribution margins of the measures were evaluated.

1.2 Structure of the dissertation

The thesis is presented as a cumulative dissertation and consists of three scientific papers. One paper is published. Two papers are submitted and currently under review.

The first paper titled "Suppressing *Alopecurus myosuroides* Huds. in Rotations of Winter-Annual and Spring Crops" is published in the MDPI open access Journal "Agriculture". It describes the impact of preventive, cultural and chemical IWM strategies on *A. myosuroides* abundance, herbicide efficacy and crop yield.

The second paper titled "A long-term study of different crop rotations and herbicide strategies: Effects on *Alopecurus myosuroides Huds.* abundance and resistance development" is submitted to ELSEVIER Journal "Crop Protection". It investigates the effect of preventive, cultural and chemical IWM strategies on crop yield, herbicide efficacy and *A. myosuroides* abundance as well as resistance development.

The third paper titled "A long-term study of crop rotations, herbicide strategies and tillage practices: Effects on *Alopecurus myosuroides* Huds. abundance and contribution margins of the cropping systems" is submitted to ELSEVIER Journal "Crop Protection". It illustrates the effect of preventive, cultural, chemical and mechanical IWM strategies on *A. myosuroides* abundance, herbicide efficacy and crop yield. Further, variable costs and contribution margins were investigated to evaluate the IWM strategies.

The format and citation style of the papers, which are presented in this thesis have been formatted uniformly.

Chapter II

—

Suppressing *Alopecurus myosuroides* Huds. in Rotations of Winter-Annual and Spring Crops

Alexander K. Zeller[a], Yasmin I. Kaiser[a] and Roland Gerhards[a]

[a] *Department of Weed Science, Institute of Phytomedicine, University of Hohenheim, 70599 Stuttgart, Germany*

Published in: *Agriculture* (2018)

8 (7), 1 – 10

doi:10.3390/agriculture8070091

2 Suppressing *Alopecurus myosuroides* Huds. in Rotations of Winter-Annual and Spring Crops

2.1 Abstract

Alopecurus myosuroides Huds. has become one of the most abundant grass weeds in Europe. High percentages of winter-annual crops in the rotation, earlier sowing of winter wheat and non-inversion tillage favor *A. myosuroides*. Additionally, many populations in Europe have developed resistance to acetyl-CoA carboxylase (ACCase), acetolactate synthase (ALS) and photosynthetic (PSII) inhibitors. Hence, yield losses due to *A. myosuroides* have increased. On-farm studies have been carried out in Southern Germany over five years to investigate abundance, control efficacies and crop yield losses due to *A. myosuroides*. Three crop rotations were established with varying proportions of winter- and summer-annual crops. The crop rotations had a share of 0, 25 and 50% of summer-annual crops. Within each crop rotation, three herbicide strategies were tested. In contrast to classical herbicidal mixtures and sequences, the aim of one of the herbicide strategies was to keep selection pressure as low as possible by using each mode of action (MOA) only once during the five years. *A. myosuroides* population was susceptible to all herbicide at the beginning of the experiment. Initial average density was 14 plants m^{-2}. In the rotation with only winter-annual crops, density increased to 5347 ears m^{-2} in the untreated control plots. Densities were lower in the rotations with 25% and even lower with 50% summer-annual crops. Control efficacies against *A. myosuroides* in the herbicide strategy using only MOAs of the HRAC-groups B and A, according to the Herbicide Resistance Action Committee (HRAC) classification on MOA, dropped after five years compared to the strategy of changing MOA in every year. Nevertheless, the results demonstrate the need for combining preventive and direct weed-management strategies to suppress *A. myosuroides* and maintain high weed-control efficacies of the herbicides.

Keywords: mode of action (MOA); preventive weed control; herbicide resistance management; crop rotation; herbicide rotation

2.2 Introduction

Crop rotations can be very effective at controlling weeds in Integrated Weed Management (IWM) (Swanton and Murphy, 1996). However, crop diversity has decreased by 50–70% in European cropping systems within the past 50 years. This is due to the use of synthetic fertilizers and pesticides (Walker and Buchanan, 1982). Production of spring cereals, potatoes, fiber crops and legumes has decreased, whereas production of winter cereals, oilseed rape and corn has increased. Winter wheat, winter barley and winter oilseed rape are dominant in moderate and humid areas with often 75–100% winter-annual crops in the rotations (Gerhards et al., 2013). Winter cereals realizes higher yield output than spring cereals and achieve higher contribution margins (OECD, 2018). The combination of cost reduction due to a minimized cultivation and a herbicide-related system used as described by Power and Follet (1987), made the system sustainable.

Alopecurus myosuroides Huds. is a winter-annual weed predominantly germinating in autumn (Naylor, 1972a). It prefers heavy, loamy, and waterlogged soils. In Western Europe, *A. myosuroides* has become very abundant in winter wheat and winter oilseed rape, particularly in early sown winter cereals after reduced tillage practices (Moss, 1987a; Melander, 1995; Lutman et al., 2013). *A. myosuroides* produces about 100 seeds per ear with a lifetime of up to 10 years (Moss, 1987a). It is a very competitive grass weed in winter wheat with 100 plants m^{-2} resulting in crop yield losses of approximately 20% (Moss, 1987b; Blair et al., 1999; Moss, 2017). Infestation rates of 500 plants m^{-2} cause yield losses of up to 50% (Moss, 1987b; Blair et al., 1999; Lutman et al., 2013; Moss, 2017).

Due to continuous applications of herbicides with the same modes of action (MOA), there has been a selection for herbicide-resistant weed populations (Powles and

Yu, 2010). Because of widespread evolved resistances, a lot of *A. myosuroides* populations have survived standard herbicide applications. Therefore, *A. myosuroides* has become the most problematic weed species in Europe (Moss, 2017). Populations with evolved resistance to herbicides have been documented in almost all European countries. Resistances to ACCase-, ALS- and PS2- inhibitors are widespread in Germany (Drobny et al., 2006). Several *A. myosuroides* populations showed cross- and multiple-resistances (De Prado and Franco, 2004). Nevertheless, farmers prefer cultivating winter wheat because it provides higher contribution margins than spring cereals (Gerhards et al., 2016). They usually start resistance management once the problem has become very evident.

There are studies that highlight the influence of spring barley on *A. myosuroides* densities (Blair, 1999; Lutman et al., 2013; Freckleton et al., 2018). Furthermore, recent studies have investigated the effect of herbicide mixtures and sequences, which are intended to prevent resistance development by using different MOA (Hicks et al., 2018). In our five-year study, we demonstrate the long-term effect on *A. myosuroides* densities by summer-annual crops, and the influence of different proportions of summer-annual crops in the crop rotation. Furthermore, we show the combination and interaction between different herbicide strategies. Additionally, we deviate from typical herbicide mixtures and sequences by setting the focus on minimal selection pressure, and not on the herbicide efficacy of *A. myosuroides* as usual.

We tested, (1) how much summer-annual crops in a rotation can reduce *A. myosuroides* densities compared to typical winter-annual cropping systems under conditions in Southern Germany. For this purpose, three different crop rotations were carried out, which differed in their proportion of summer-annual crops (0, 25 and 50%). The second hypothesis was (2), that using every herbicide MOA only once over the five-year period would result in higher *A. myosuroides* control efficacies than using herbicides of the HRAC-groups B and A every year, in the long term. Finally (3), we assumed that only through the combination of preventive measures – crop rotation – and minimal selection pressure – annual MOA change – could a satisfactory result be achieved.

2.3 Material and Methods

2.3.1 Field Experiment and Trial Design

The experiment was designed as a randomized split-plot. The main plot factor was crop rotation (CR) and the sub-plot factor herbicide strategy (HS). The size of each sub-plot was 6 m×12 m. Each combination of CR×HS was replicated 4 times. In CR1, 3 years of winter wheat (*Triticum aestivum* L.) was followed by winter oilseed rape (*Brassica napus* L.). In CR2, winter wheat in the 3rd year was replaced by spring barley (*Hordeum vulgare* L.) and in CR3, winter wheat in the 2nd year was replaced by corn (*Zea mays* L.), and spring barley was grown in the 3rd year. In Year 5, crop rotation started again with winter wheat in all variances. The experimental design was set up to test different proportions of summer-annual crops in the rotations (CR1 0%, CR2 25% and CR3 50%) on the infestation of *A. myosuroides*. However, with this experimental design, the results of the experiment are correlated with the annual conditions. Therefore, it was not possible to capture the difference between crop and year separately. Three weed-control strategies were included in each crop and year. HS1 was the untreated control. Only broad-leaved weeds were controlled. In HS2, herbicide MOA was changed in every year to reduce the selection pressure on a single MOA. Therefore, each MOA was used only once over the 5-year experiment. To ensure this, active ingredients were also used which are known to have only weak efficacies to *A. myosuroides* or high dependence on weather conditions.

In HS3, the selection pressure was high, and we tried to use only herbicides of HRAC-group B. If this was not possible due to the cultivated crop (spring barley and winter oilseed rape), herbicides of HRAC-group A were used, which are also among the highly effective but also high resistance-risk MOA (Table 1). Broad-leaved weeds were controlled in all herbicide strategies with active ingredients that had no activity against grasses.

Table 1. Crop rotations (CR), sowing dates and herbicide strategies (HS) of the experiment.

Crop Rotation Sowing Date	Year				
	1	2	3	4	5
1	winter wheat 25/10/2011	winter wheat 31/10/2012	winter wheat 19/11/2013	winter oilseed rape 23/08/2014	winter wheat 12/10/2015
2	winter wheat 25/10/2011	winter wheat 31/10/2012	summer barley 19/03/2014	winter oilseed rape 23/08/2014	winter wheat 12/10/2015
3	winter wheat 25/10/2011	maize 25/04/2013	summer barley 19/03/2014	winter oilseed rape 23/08/2014	winter wheat 12/10/2015
	Herbicide strategy (HS)				
1	untreated control *				
2	change of herbicide MOA every year				
3	herbicides of the HRAC-groups B and A				

* only broad-leaved weeds were controlled.

The field experiment took place in South-west Germany (48.74◦ N, 8.92◦ E, 478 m altitude). The study started in autumn 2011 and continued until summer 2016. In 2011, pea was cultivated before winter wheat. After 3 passes of chisel plough, winter wheat was sown. The average annual temperature is 9.1 °C and the average annual precipitation is 825mm. The soil texture is a clayey loam. The cultivars Schamane (Saatzucht Streng, Uffenheim, Germany; winter wheat), Torres (KWS, Saat AG, Einbeck, Germany; maize), Grace (Baywa, München, Germany; spring barley) and Avatar (Rapool, Isernhagen HB, Germany; oilseed rape) were sown in the experiment. Sowing dates were adapted to the weather conditions (Table 1). For cereals and oilseed rape, a single disc drill (row distance 0.12 m) and for corn a precision air seeder (row distance 0.75 m) were used. Prior to the spring-crops, lacy phacelia (*Phacelia tanacetifolia* Benth.) was cultivated as a cover crop (cv. Julia, Feldsaaten Freudenberger, Magdeburg, Germany). Glyphosate (1440 g a.i. ha^{-1}, Clinic®, 360 g a.i. L^{-1}, Nufarm Deutschland GmbH, Cologne, Germany) was sprayed to remove over-wintering cover crops, volunteer cereals and weeds. Subsequently, shallow soil tillage was carried out and the spring crops were sown. All crops were fertilized as usual. Fungicides and insecticides were applied if necessary. Throughout the five-year study, only non-inversion tillage operations (typical for south-west Germany) not deeper than 0.12 m were performed. Herbicides were applied with a self-propelled plot sprayer (Schachtner-Gerätetechnik, Ludwigsburg, Germany), which was calibrated for a volume of 200 L ha^{-1}. Crop yield was recorded in 4 × 6 m sub-plots using a plot combine harvester. All herbicides were applied at the recommended dose (Table 2).

Table 2. Herbicides applied in different crops and years to control *A. myosuroides*.

Year	Crop	HS	Application Time	Herbicide (Trade Name)	Active Ingredient	HRAC-Code	Rate (g a.i. ha^{-1})
1	WW	2 & 3	spring	Broadway® + Adj. [1]	pyroxsulam + florasulam *	B	15 + 5 *
2	WW	2	spring	Arelon® TOP [3]	isoproturon	C1	1500
		3	spring	Broadway® + Adj. [1]	pyroxsulam + florasulam *	B	15 + 5 *
	C	2	spring	Laudis®[2]	tembotrione	F2	88
		3	spring	Elumis®[5]	mesotrione * + nicosulfuron	F2 * + B	93.8 * + 37.5
3	WW	2	autumn; spring	Herold® SC [6]; Traxos®[5]	flufenacet + diflufenican; clodinafop + pinoxaden	K3 + F1; A	240 + 120; 30 + 30
		3	spring	Broadway® + Adj. [1]	pyroxsulam + florasulam *	B	15 + 5 *
	SB	2	spring	Axial 50®[5]	pinoxaden	A	60
		3	spring	Axial 50®[5]	pinoxaden	A	60
4	OR	2	autumn; winter	Butisan® Gold + Stomp® Aqua [4]; Kerb Flo®[1]	dimethenamid * + metazachlor + pendimethalin; propyzamid	K3 * + K3 + K1; K1	400 * + 400 + 341,25; 750
		3	autumn	Gallant® Super [1]	haloxyfop	A	52
5	WW	2	autumn	Boxer®[5]	prosulfocarb	N	4000
		3	spring	Broadway® + Adj. [1]	pyroxsulam + florasulam *	B	15 + 5 *

Adj. = adjuvant; WW = winter wheat; C = corn; SB = summer barley; OR = winter oilseed rape; * = No efficacy against *A. myosuroides*. [1] DOW AgroSciences GmbH, Munich, Germany; [2] Bayer AG CropScience, Monheim am Rhein, Germany; [3] FMC Corporation Cheminova Deutschland GmbH, Stade, Germany; [4] BASF Plant Protection, Ludwigshafen, Germany; [5] Syngenta Agro GmbH, Maintal, Germany.

2.3.2 Assessments of *A. myosuroides* Density and Control Efficacy

The effects of CR and HS on *A. myosuroides* densities were assessed by counting ears of *A. myosuroides* in 5 randomly placed frames (0.4 m^2) per plot in each crop and year at flowering stage of the crop. *A. myosuroides* control efficacy (ACE) was calculated using Equation (1):

$$\text{ACE (\%)} = \left[\frac{\text{A}-\text{B}}{\text{A}}\right] * 100 \quad (1)$$

Equation 1. *A. myosuroides* control efficacy (ACE)

where A represents the number of *A. myosuroides* ears m^{-2} in the untreated control plots and B represents the number of *A. myosuroides* ears m^{-2} in the treated HS plots 14 days after treatment.

2.3.3 Statistical Analysis

Data analysis was performed with the statistical software R® 3.3.3 (R Development Core Team, Vienna, Austria, 2011). A linear mixed-effect model was used to evaluate the response of *A. myosuroides* to CR and HS as fixed factors. A covariance matrix θ was applied during the analysis of data. Results were log transformed to homogenize variances and to normalize the distribution. In the results section, back transformed means are shown. Crop yield (t ha^{-1}) and ACE (%) were evaluated in each crop and year separately. Yield was analyzed by ANOVA and ACE by linear mixed model. Multiple mean comparison tests were performed using the Tukey test at a significance level of $\alpha \leq 0.05$. Figures were created with Sigmaplot V12.5 (Systat software GmbH, Erkrath, Germany).

2.4 Results

Crop rotation (CR), herbicide strategy (HS) and year, as well as their interaction, significantly influenced *A. myosuroides* density. *A. myosuroides* density was relatively low at the beginning of the experiment with an average density of 14 ears m^{-2}. Plants were distributed homogenously within the experiment at the beginning. Infestation rates were below or close to the economic weed thresholds of 10–15 plants m^{-2} (Gerowitt and Heitefuss, 1990). During the study, densities significantly increased in all three rotations in the untreated control plots (HS1) (Figures 1 and 2, and Table 3). In the third year of winter wheat in CR1, *A. myosuroides* density increased to 883 ears m^{-2}. After five years, a maximum density of 5347 ears m^{-2} was counted in CR1 (Figure 1). When winter wheat was replaced by spring barley in the third year (CR2), densities of *A. myosuroides* were

33% lower than in CR1. In CR3, with maize and spring barley, *A. myosuroides* densities were 50% lower than in CR1, after five years (Figure 1, Table 3).

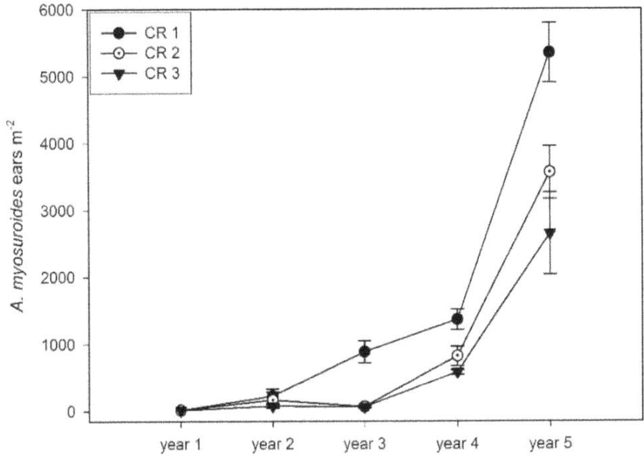

Figure 1. *A. myosuroides* ears m^{-2} in the untreated control plots of crop rotations (CR) 1, 2 and 3 over the five-year experiment at Ihinger Hof, Renningen, Germany.

Herbicide strategies also had a strong impact on *A. myosuroides* densities (Figure 2). In the first three years of the experiment, densities and control efficacies were similar in the strategies using MOA of HRAC-groups B and A and different MOA every year. Weed-control efficacies, weed densities and crop yields were only significantly different from the untreated control. In the second year, ACE ranged between 85%–95% in winter wheat. In maize, HS3 showed the best ACE with 100%, whereas HS2 performed significantly worse with 77%. The reason for the low control efficacy in HS2 was due to the use of tembotrione, which is less active against *A. myosuroides* than nicosulfuron in HS3. Nevertheless, it was applied in this experiment in HS2 to ensure a continuous change of herbicide MOA. *A. myosuroides* infestations in the untreated control caused tremendous yield losses of 81% compared to both herbicide strategies.

In contrast to maize, spring barley yield was not reduced by *A. myosuroides* competition. In the third year, spring barley yields were equal in all herbicide strategies and not different to the untreated control, although 65 *A. myosuroides* ears m^{-2} were counted in the control plots. *A. myosuroides* infestation in the third year of winter wheat in CR1 reduced grain yield by 39% compared to HS2 and HS3 (Table 3, Figure 2). Over the period of five years, herbicide efficacy was highest in HS2 (continuous change of herbicide MOA), followed by HS3. From the fourth year onwards, ACE of HS3, using exclusively HRAC- groups B and A herbicides, was significantly lower than in HS2. This resulted in a rapid increase of *A. myosuroides* densities in HS3 in the fifth year of study. Surprisingly, application of only the pre-emergent herbicide prosulfocarb in HS2 resulted in 99% control efficacy against *A. myosuroides*. No post-emergent herbicide was needed in spring. This supports the observations of Naylor (1972b) that seeds of *A. myosuroides* mostly germinate in autumn. However, efficacy of pre-emergent herbicides could be much lower when the soil is dry after seeding winter-annual crops (Melander, 1995).

Table 3. Average densities of *A. myosuroides* ears m^{-2}, *A. myosuroides* control efficacy (ACE) and crop yield in the crop rotations (CR) and herbicide strategies (HS).

Year	CR	HS	*A. myosuroides* Ears m^{-2}	Standard Errors	ACE (%)	Crop Yield (t ha^{-1})
1	1; 2; 3—WW	1; 2; 3	14	1.6	99	6.93
2	1—WW	1	222 a	94	—	7.3 a
		2	20 b	14	91 a	8.6 a
		3	12 b	7	95 a	8.4 a
	2—WW	1	166 a	105	—	7.9 a
		2	25 b	14	85 a	8.8 a
		3	16 b	7	90 a	8.2 a
	3—M	1	75 a	20	—	1.5 b
		2	18 b	3	77 b	6.9 a
		3	0 c	0	100 a	7.0 a
3	1—WW	1	883 a	140	—	5.6 b
		2	33 b	25	96 a	9.0 a
		3	6 c	9	99 a	9.3 a
	2—SB	1	65 a	14	—	6.7 a
		2	0 b	0	100 a	6.8 a
		3	1 b	1	99 a	7.0 a
	3—SB	1	63 a	15	—	6.5 a
		2	1 b	0.6	99 a	7.4 a
		3	0 b	0	100 a	6.0 a
4	1—OR	1	1365 a	133	—	0.7 b
		2	1 c	0.6	100 a	5.0 a
		3	81 b	25	94 b	4.4 a
	2—OR	1	822 a	123	—	1.4 b
		2	0 c	0	100 a	4.7 a
		3	56 b	71	93 b	4.6 a
	3—OR	1	582 a	28	—	1.1 b
		2	0 c	0	100 a	4.0 a
		3	44 b	39	93 b	3.7 a
5	1—WW	1	5347 a	381	—	0.8 b
		2	120 c	6	98 a	5.8 a
		3	917 b	466	83 b	4.6 ab
	2—WW	1	3562 a	340	—	0.9 b
		2	77 c	17	98 a	6.0 a
		3	770 b	307	78 b	4.3 ab
	3—WW	1	2648 a	230	—	0.9 b
		2	39 c	6	99 a	5.7 a
		3	366 b	124	86 b	4.1 ab

WW= winter wheat; M =maize, SB = spring barley, OR = winter oilseed rape; Means followed by the same letter are not significantly different according to Tukey's HSD method at the P = 0.05 significance level.

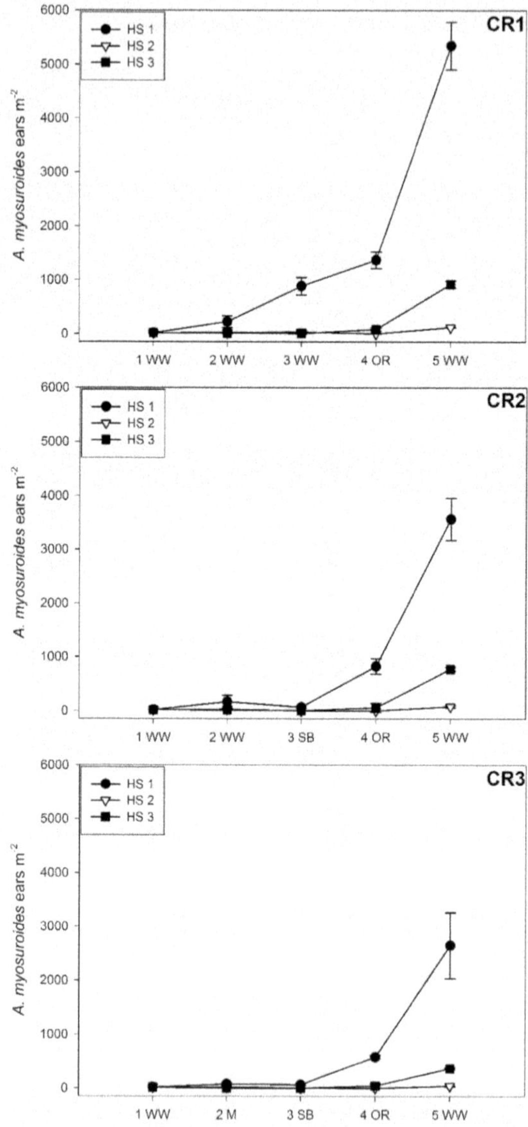

Figure 2. *A. myosuroides* densities expressed in ears m^{-2} at the different herbicide strategies (HS) in crop rotations (CR).

2.5 Discussion

Alopecurus myosuroides is a very competitive weed species that is well adapted to winter wheat production in Europe. It can rapidly increase population density when no efficient control methods are applied and rotations with a high percentage of winter-annual crops and reduced tillage are practiced. This was shown in this study and it agrees with the results of References (Moss, 1987a, 1987b; Melander, 1995; Moss, 2017; Freckleton et al., 2018). Yields of oilseed rape (Year 4) and winter wheat (Year 5) in the untreated plots were approximately 85% lower than in the most effective herbicide treatments (Table 3). These results are in line with Blair et al. (1999), who also observed 80% grain yield losses due to *A. myosuroides* competition in winter wheat.

Lutman et al. (2013) reported that spring barley reduced *A. myosuroides* densities on average by 88%, because the majority of *A. myosuroides* seeds germinate in autumn (Naylor, 1972b; Moss, 1990). In our study, the inclusion of summer-annual crops reduced *A. myosuroides* densities by 61% in Year 2 (CR3) and 93% (CR2 + 3) in Year 3. After five years, a proportion of 25% summer-annual crops in the rotation (CR2) reduced *A. myosuroides* densities by 33%. With a proportion of 50%, when spring barley and maize were included (CR3) a reduction of 50% could be achieved. Freckleton et al. (2018) were also able to demonstrate the positive influence of spring barley on the reduction of *A. myosuroides* densities. They assume that in addition to the main germination in autumn, the tillage and seedbed preparation in the spring remove most of the plants. Secondly, they cite the competitiveness of spring barley, which is characterized by rapid growth and high biomass production, which can suppress the development of *A. myosuroides* effectively. This coincides with our results. We were able to achieve a significant *A. myosuroides* reduction with corn, but only with spring barley yield could be maintained in HS1. Late seeding of winter wheat, rotational ploughing, false seedbed preparation, increased crop density and growing competitive winter wheat cultivars are also efficient preventive methods to suppress *A. myosuroides* (Lutman et al., 2013).

Before the start of the experiment, a greenhouse bioassay with 10 different herbicides of the active ingredient classes HRAC-groups A, B and C was performed. For all herbicides, efficacies between 95–100% were estimated. Therefore, the population could be classified as susceptible (data not shown). In the last two years of the experiment (Years 4 and 5), weed-control efficacy decreased, when only HRAC-groups B and A herbicides were used compared to herbicide strategy rotating MOA every year. The weed population in HS3 shifted from a sensitive to a resistant population, when exclusively herbicides with the same MOA (B and A) were applied. During the first three years, however, efficacy of herbicides in HS3 remained very high. Gressel and Segel (1990) came to similar results for maize and triazine-resistant weeds. Neve and Powles (2005) selected for resistant *Lolium rigidum* plants with repeated low-dose applications of ACCase-inhibitors over five generations. With the resistance coming up in HS3, densities of *A. myosuroides* increased very fast, especially in CR1 with only winter-annual crops (Table 3). Hicks et al. (2018) also reported the possibility of a fast resistance development and could show a significant correlation of resistance development and high *A. myosuroides* densities. This is also consistent with our results, which showed that the more winter-annual crops used inside the crop rotation, the higher the proportion of resistant plants.

This experiment clearly underlines the need for combining different weed-control methods. Rotating winter- and summer-annual crops effectively suppresses weed species that predominantly germinate in spring or autumn such as *A. myosuroides* (Lutman et al., 2013; Freckleton et al., 2018). The lower densities lead to fewer mutations – which naturally occur in a population – and therefore to a lower selection by herbicides. In addition, various crops also allow a wider range of active ingredients to be used. Moreover, if farmers avoid the repeated use of the same active ingredients for as long as possible, a development of resistance can be prevented, although this may lead to temporarily higher densities, especially when using weaker or weather-dependent active ingredients. However, the higher infestation often has no negative effect on the yield, as our results, and the results of Hicks et al. (2018) have demonstrated. Hicks et al. (2018) could detect yield losses between 2–12%. Nevertheless, significant losses occurred only at high and very high densities.

We are aware that the results of this experiment are limited. Geographic differences such as soil type or climate could influence the results shown. Furthermore, the effect of crop and year, which we were not able to distinguish, may have led to fluctuations in *A. myosuroides* density due to weather conditions. Nevertheless, we believe that such accurate long-term studies are important and rare. Studies like this show clear trends from which working management methods can be derived. In the near future, no new active ingredients are expected. An increasing number of active ingredients lost or might lose their application permission during the re-registration process (Duke, 2011). Therefore, the available active ingredients must be used sustainably and maintained for as long as possible. Only in this way can consistent yields be guaranteed in the future.

However, autumn and spring cropping is restricted to areas with moderate temperatures during the winter. In many parts of the world, only spring cropping is possible. For those areas, several other preventive methods of non-chemical weed control are possible, such as false seedbed preparation, rotational ploughing, intercropping, delayed sowing, higher seed rates and competitive crop cultivars (Knab and Hurle, 1988; Lutman et al., 2013; Gerhards et al., 2016).

2.5.1 Author Contributions:

R.G. and Y.I.K. conceived and designed the experiments; A.K.Z. and Y.I.K. performed the experiments and analyzed the data; A.K.Z. contributed to the materials and analysis tools; R.G. and A.K.Z. wrote the paper.

2.5.2 Funding:

This study was funded by the German Ministry of Agriculture and the German Rentenbank.

2.5.3 Acknowledgments:

The authors thank Christoph Gutjahr for initiating the study. We appreciate the technical assistance at the Research Station Ihinger Hof during the field trials. We thank Alexandra Heyn for her help with the greenhouse biotests. The authors thank the German Ministry of Agriculture and the German Rentenbank for funding this study.

2.5.4 Conflicts of Interest:

The authors declare no conflicts of interest.

Chapter III

—

A long-term study of different crop rotations and herbicide strategies: Effects on *Alopecurus myosuroides Huds.* abundance and resistance development

Alexander K. Zeller[a], Yasmin I. Zeller[b] and Roland Gerhards[a]

[a]*University of Hohenheim, Institute of Phytomedicine, Department of Weed-Science, Otto-Sander-Str. 5, 70599 Stuttgart, Germany*

[b]*FMC Agricultural Solutions, Westhafenplatz 1, 60327 Frankfurt am Main, Germany; previously University of Hohenheim*

Submitted to: *Crop Protection*

3 A long-term study of different crop rotations and herbicide strategies: Effects on *Alopecurus myosuroides Huds.* abundance and resistance development

3.1 Abstract

Alopecurus myosuroides Huds. (blackgrass) is considered as one of the most problematic weeds in Europe. Changes in the agricultural practice such as high proportions of winter-annual crops in the crop rotation, early sowing dates of cereals and non-inversion tillage systems promoted blackgrass infestations in the last decades. The intensive use of herbicides led to the development of herbicide resistant blackgrass populations in many European countries. Five year on-farm studies were conducted at two locations in Southwestern Germany, to investigate the impact of crop rotations and herbicide strategies on blackgrass and resistance development. Crop rotations with 100 %, 75 % and 50 % winter annual crops were established as main factor and four herbicide strategies with different selection pressure on resistance development were included as sub-plot factor in each crop. Blackgrass densities increased from 14 to 5570 and 201 to 3546 heads m^{-2} at the two experimental sites, in the untreated control when exclusively winter annual crops were cultivated. Herbicide efficacies decreased by up to 55 % when equal herbicide modes of action were used in every year. Only the rotation with 50 % winter annual crops and a rotation of herbicide modes of action in every year prevented an increase of blackgrass density and herbicide resistance development. These results highlight the need for integrated weed management with multiple strategies to control problematic weed species such as blackgrass.

Keywords: blackgrass, resistance management, integrated weed management, mode of action, herbicide mixtures

3.2 Introduction

Traditionally wide crop rotations have been utilized to control weeds and pests in temperate areas (Leighty, 1938). With the opportunity to use synthetic fertilizers and pesticides, crop rotations have been simplified and tillage practices reduced (Walker and Buchanan, 1982; Power and Follet, 1987). Blackgrass, *Alopecurus myosuroides* Huds. (Poales: Poaceae), is a winter-annual grass which mainly germinates in autumn (Naylor, 1972a; Moss, 1985). The propagation of blackgrass was favored by crop rotations with more than 75 % of winter wheat (*Triticum aestivum* L.), winter barley (*Hordeum vulgaris* L.) and winter oilseed rape (*Brassica napus* L.) (Gerhards et al., 2013). Blackgrass plants can produce up to 1200 viable seeds plant^{-1} and arise mainly from fresh seeds not older than 1 year in non-inversion tillage systems (Naylor, 1972b). Populations can rapidly increase by 10-fold per annum when no effective weed control management is performed (Moss, 1987a). Naylor (1972a) and Blair et al. (1999) reported yield losses of 13, 32, 37 and 50% in winter wheat due to blackgrass infestations of 30, 100, 300 and 400 plants m^{-2}, respectively. Results of Fritzsche et al. (2012) have shown that infestations of 10 heads m^{-2} result in yield losses of 0.05 t ha^{-1} each in winter cereals.

The use of herbicides is seen as the most effective and economical weed control method and replaced almost all other control strategies, since their introduction into the market (Heap, 2014). However, high abundances, genetic diversity and the repeated herbicide use with the same modes of action (MOAs) have shifted populations towards resistant biotypes (Powles and Yu, 2010). As a result of cross-pollination, resistances to different MOAs and chemical families can accumulate in one biotype (Werck-Reichhart et al., 2000; De Prado and Franco, 2004). Cross- and multiple-resistant populations are reported throughout Europe (Heap, 2019). Drobny (2016) reported a decreasing number of available active ingredients (AIs) to control weeds due to authorization losses in the course of the re-registration process. Several AIs for blackgrass control including fenoxaprop (Germany), isoproturon and flupyrsulfuron (Europe) are no longer available. By contrast, only one new AI with efficacy on blackgrass is supposed to enter the market (Campe et al., 2018) and there are plenty of indications that the high potential MOAs had already been discovered (Duke, 2011).

Blackgrass and other problematic weed species require integrated weed management (IWM) strategies to maintain the efficacy of existing AIs as long as possible. Recommended strategies include changes in crop rotation and an alternation of herbicide MOA by herbicide-sequences or -mixtures (Chauvel et al., 2001; Beckie, 2006; Beckie and Reboud, 2009; Moss et al., 2007; Lagator et al., 2013). The aim of this five-year field study at two locations was to investigate blackgrass populations with regard to control efficacy, resistance development and its impact on crop yield affected by three crop rotations (CR), four herbicide strategies (HS) and their interaction. It was hypothesized that a) in diverse CRs blackgrass densities are lower, crop yield is higher and herbicide resistance development is slower than in CRs with only winter annual crops. Further, we assumed that b) a rotating HS, which uses each MOA or at least chemical family only once during the experiment will perform better than herbicide-sequences and -mixtures. Finally, we hypothesized that c) only a combination of a diverse CR and herbicide rotation can decelerate herbicide resistance development.

3.3 Materials and methods

3.3.1 Field experiments

The field experiments were located in Southwestern Germany and started in autumn 2011 at Ihinger Hof (Renningen; 48.74° N, 8.92° E, 478 m ASL) and 2012 at Wurmberg (48.86° N, 8.83 °E, 450 m ASL). The average annual temperature and precipitation was 9.1°C and 825 mm at Ihinger Hof and 10.6 °C and 736 mm at Wurmberg. The soil type was a clayey loam at both locations. The field experiments were designed as randomized split-plots with four replicates. The main plot factor was crop rotation (CR) with a plot size of 24 x 24 m. Three different CRs were tested with proportions of 100 %, 75 % and 50 % winter-annual crops (Table 4). All CRs had a duration of four years. The experiment ended, when winter wheat was grown again in all treatments in the fifth year.

Table 4. Crop rotations (CR) and sowing dates of each crop and location.

CR / sowing dates	Year 1	2	3	4	5
CR 1	WW	WW	WW	WOR	WW
Ihinger Hof	25/10/2011	31/10/2012	19/11/2013	23/08/2014	12/10/2015
Wurmberg	03/10/2012	04/10/2013	08/10/2014	21/08/2015	15/10/2016
CR 2	WW	WW	SB	WOR	WW
Ihinger Hof	25/10/2011	31/10/2012	19/03/2014	23/08/2014	12/10/2015
Wurmberg	03/10/2012	04/10/2013	18/03/2015	21/08/2015	15/10/2016
CR 3	WW	C	SB	WOR	WW
Ihinger Hof	25/10/2011	25/04/2013	19/03/2014	23/08/2014	12/10/2015
Wurmberg	03/10/2012	16/04/2014	18/03/2015	21/08/2015	15/10/2016

CR = crop rotation, WW = winter wheat, WOR = winter oilseed rape, SB = spring barley, C = corn

The sub-plot factor was herbicide strategy (HS) and the sub-plot size 6 x 12 m. The sub-plots included an untreated control (HS1), a rotation of herbicides in every year (HS2), a treatment according to recommendations of local plant protection services (HS3) and the application of herbicides from HRAC group A or B according to the Herbicide Resistance Action Committee (HRAC) in every year (HS4). In HS4 herbicides with a high efficacy on blackgrass were used which are known to exert high selection pressure on resistance. HS3 aimed to exert a medium selection pressure performing herbicide-mixtures or -sequences. In HS2, the aim was to use every MOA or at least chemical family only once during experimental time to minimize the selection pressure as much as possible. To control dicotyledonous weeds, herbicides were used which had no blackgrass activity. All herbicides were applied at the recommended rate (Table 5) using a self-propelled plot sprayer (Schachtner - Gerätetechnik, Ludwigsburg, Germany), which was calibrated for a volume of 200 L ha^{-1} and a spray pressure of 170 kPa. Air injector double flat fan nozzles (02-120° IDTK, Lechler, Metzingen, Germany) were mounted on the spray boom. To avoid cross-pollination and to separate the 48 plots, glyphosate (1440 g AI ha^{-1}, Clinic®, Nufarm GmbH, Cologne, Germany) was sprayed on 0.4 m strips around the sub-plot.

Table 5. Herbicides applied in different crops and years to control blackgrass.

Year	Crop	Timing	HS	Tradename	Formulation	Active ingredient	HRAC - Code	Rate (g AI ha^{-1})
1	WW	Spring	2	Broadway® + Adj.[1]	WG	pyroxsulam + florasulam*	B	15 + 5*
		Spring	3	Broadway® + Adj.[1]	WG	pyroxsulam + florasulam*	B	15 + 5*
		Spring	4	Broadway® + Adj.[1]	WG	pyroxsulam + florasulam*	B	15 + 5*
2	WW	Spring	2	Arelon®[2]	SC	isoproturon	C1	1500
		Spring	3	Atlantis® + Adj.[3]	WG	mesosulfuron + iodosulfuron	B	9 + 1.8
		Spring	4	Broadway® + Adj.[1]	WG	pyroxsulam + florasulam*	B	15 + 5*
	C	Spring	2	Laudis®[3]	OD	tembotrione	F2	88
		Spring	3	Kelvin®[4]	OD	nicosulfuron	B	32
		Spring	4	Elumis®[5]	OD	mesotrione* + nicosulfuron	F2* + B	93.8* + 37.5
3	WW	autumn; spring	2	Herold®[6]; Traxos®[5]	SC EC	flufenacet + diflufenican clodinafop + pinoxaden	K3 + F1; A	240 + 120 30 + 30
		Autumn	3	Herold®[6]	SC	flufenacet + diflufenican	K3 + F1	240 + 120
		Spring	4	Broadway® + Adj.[1]	WG	pyroxsulam + florasulam*	B	15 + 5*
	SB	Spring	2	Axial® 50[5]	EC	pinoxaden	A	60
		Spring	3	Axial® 50[5]	EC	pinoxaden	A	60
		Spring	4	Axial® 50[5]	EC	pinoxaden	A	60
4	WOR	autumn; winter	2	Butisan® AquaPack[4]; Kerb™ Flo[1]	EC/CS SC	metazachlor + pendimethalin propyzamide	K3 + K1; K1	400 + 341.25 750
		Autumn	3	Butisan® Aqua-Pack[4]; Select® 240[2]	EC/CS EC	metazachlor + pendimethalin clethodim	K3 + K1; A	400 + 341.25 121
		Autumn	4	Gallant® Super[1]	EC	haloxyfop	A	52
5	WW	Autumn	2	Boxer®[5]	EC	prosulfocarb	N	4000
		Autumn	3	Lexus®[2] + Malibu®[4]	EC EC	flupyrsulfuron pendimethalin + flufenacet	B K1+K3	10 1200 + 240
		Spring	4	Broadway® + Adj.[1]	WDG	pyroxsulam + florasulam*	B	15 + 5*

HS = herbicide strategy, WW = winter wheat, C = corn, SB = spring barley, WOR = winter oilseed rape, Adj = adjuvant; * No efficacy on blackgrass, herbicides against broad-leaved weeds are not shown. [1]Corteva Agriscience™ (Munich, Germany), [2]FMC Agricultural Solutions (Stade, Germany), [3]Bayer CropScience GmbH (Langenfeld, Germany), [4]BASF SE (Ludwigshafen, Germany), [5]Syngenta Agro GmbH (Maintal, Germany), [6]Adama GmbH (Cologne, Germany)

Sowing dates were adapted to the weather conditions (Table 4). For cereals and oilseed rape, a single disc drill (750 A, John Deere, Walldorf, Germany) with a row distance of 0.12 m and for corn a precision air seeder (Maxima 2, Kuhn GmbH, Schopsdorf, Germany) with a row distance of 0.75 m was used. Prior to the spring-crops, lacy phacelia (*Phacelia tanacetifolia* Benth.) was cultivated as a cover crop. Glyphosate (1440 g AI ha^{-1}, Clinic®, Nufarm GmbH, Cologne, Germany) was sprayed to remove over-wintering cover crops, volunteer cereals and weeds. Subsequently, spring crops were sown after shallow seedbed preparation. Fertilization was carried out as needed in all crops. Fungicides and insecticides were applied if necessary. Throughout the five-year study, only non-inversion tillage operations (typical for Southwestern Germany) not deeper than 0.12 m were performed.

3.3.2 Observations and measurements

3.3.2.1 Blackgrass densities and yield

The effect of CR and HS on blackgrass was assessed by counting blackgrass heads m^{-2} in each crop and year, at BBCH 61-75 of the crop. Eight randomly placed counts (0.4 m^{-2}) per plot were conducted. The crop yield was recorded in the core (4 x 5 m) of each sub-plot using a plot combine harvester (Zürn170, Zürn Harvesting GmbH & Co. KG, Schöntal-Westernhausen, Germany) or a plot field chopper (MH500, Deutz Fahr GmbH, Lauingen, Germany).

3.3.2.2 Herbicide resistance investigations

In order to investigate the resistance levels in a greenhouse bioassay, ripe seeds of untreated blackgrass were sampled at both locations, prior to the start of the experiment in 2011/2012. Seeds were cultivated in vermiculite at a temperature of 15°C day and 5°C night with a 12-h photoperiod. At one-leaf stage, five seedlings were transplanted into 0.08 x 0.08 m Jiffy pots and sprayed at two-three leaf stage with a precision application chamber using a flat-fan nozzle (8002 EVS, Teejet® Spraying System Co., Wheaton, IL, USA). The adjustment was set for a volume of 200 L ha^{-1}, a speed of 0.7 m s^{-1}, a spraying pressure of 300 kPa and a distance of the spray nozzle of 0.5 m above sprayed surface. Four herbicides were tested and sprayed at the recommended rate (Table 6). Each treatment was repeated four times and pots were placed completely randomised in the cabin. Herbicide efficacy was estimated visually based on leaf area reduction, leaf discoloration and growth abnormalities (Horowitz, 1975). At the end of the fifth year, blackgrass seeds were collected in all CRs and HSs at both locations. Twelve populations per location (3 CR x 4 HS) were tested in a greenhouse bioassay under same conditions as described above. The resistance development of a population was determined using the R-Scaling method described by Moss *et al.* (1999). Herbicide efficacies of ≥ 95% were classified as susceptible (S), ≥ 76-95% (R?) were interpreted as early indications of resistance development, possibly reducing herbicide performance. Herbicide efficacies of ≥ 38-76% (RR) were rated as resistant, probably reducing herbicide performance and with efficacies of ≥ 0-38% (RRR) it was concluded that resistance was confirmed and will highly likely reduce herbicide performance.

Table 6. Herbicide treatments in the greenhouse bioassays before and in the end of the experiment.

MOA	Trade name	Formulation	AI	Rate (g AI ha^{-1})
HRAC A	Axial® 50 [1]	EC	pinoxaden	60
HRAC B	Atlantis® WG [2] + Adj.	WG	mesosulfuron + iodosulfuron	12 + 2.4
	Broadway® + Adj. [3]	WG	florasulam* + pyroxsulam	5* + 15
HRAC C	Arelon Top® [4]	SC	isoproturon	750

AI = active ingredient; Adj = adjuvant. *No efficacy against blackgrass. [1]Syngenta Agro GmbH (Maintal, Germany). [2]Bayer CropScience GmbH (Langenfeld, Germany). [3]Corteva Agriscience™ (Munich, Germany). [4]FMC Agricultural Solutions (Stade, Germany).

To determine whether the blackgrass plants survived the herbicide treatment due to mutations in the target site, leaf samples of surviving plants were collected 28 days after application for molecular genetic analysis. In total 41 leaf samples of plant material originating from Ihinger Hof (23 ACCase, 18 ALS) and 103 leaf samples from plants originating from Wurmberg (46 ACCase, 58 ALS) were tested. The most common gene loci in ACCase 1781, 2027, 2041, 2078, 2096, ALS 197, 574 and PSII 219, 220, 251, 255, 256, 264, 265, 266, 275 were examined. All polymerase chain reaction (PCR) assays and pyrosequencing procedures were conducted by IDENTXX GmbH (Stuttgart,

Germany). For each target site, a separate PCR was performed containing 50 ng of genomic DNA in each reaction and using specific primers (this information is subject to the trade secret of IDENTXX GmbH). The melting was set up at 94 °C for 5 min, the primer annealing depends on a primer specific temperature and the extending was set up at 72 °C for 5 min.

3.3.3 Statistical analysis

Data analysis was performed with the statistics software SAS® (V9.4, SAS Institute Inc, NC, USA). A linear mixed effect model was used to evaluate the response of blackgrass and crop yield with regard to the factors CR, HS and year. A square root transformation was performed for blackgrass and a log transformation for crop yield data to homogenize variances and to normalize distribution. A comparison of means was performed using the Tukey test at a significance level of $\alpha \leq 0.05$. To find a letters display for all pairwise comparisons, a SAS macro was used described by Piepho (2004; 2012). To present the data the back-transformed means were used. Both locations were analyzed separately because of the different blackgrass densities and resistance levels at the beginning of the experiment as well as the different starting times. Figures were created with Sigmaplot V12.5 (Systat software GmbH, Germany).

3.4 Results

3.4.1 Effect of crop rotations on blackgrass densities

At the beginning of the study blackgrass densities of 14 (±1.6) heads m^{-2} were counted at Ihinger Hof and 201 (+125) at Wurmberg. Blackgrass densities increased rapidly at both locations when no herbicides were used (HS1). After 5 years, blackgrass densities in CR1 have increased to 5347 heads m^{-2} at Ihinger Hof and to 3424 at Wurmberg. In CR2 and CR3, blackgrass densities were significantly lower. Especially due to the cultivation of spring barley in the third year, infestation rates were significantly reduced at both locations. The more summer annual crops were cultivated, the lower densities were observed. Thus, after five years blackgrass infestations were 33 % lower through the cultivation of one (CR2) and by 50 % lower through the cultivation of two spring crops (CR3) at Ihinger Hof, and by 7 % and 32 % at Wurmberg, respectively (Figure 3).

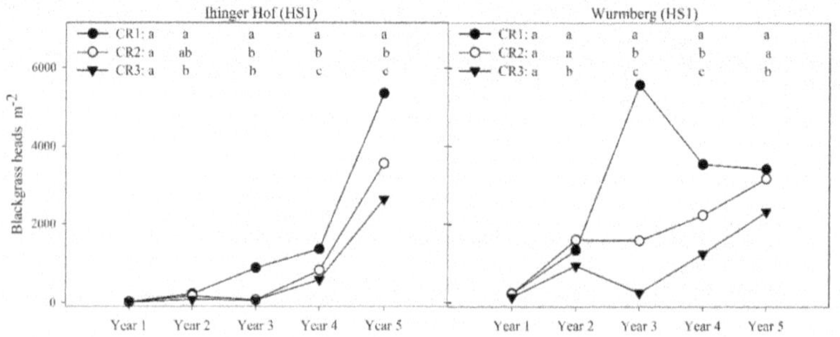

Figure 3. Blackgrass heads m^{-2} in the three crop rotations (CR) in the untreated control (HS1) at Ihinger Hof and Wurmberg. Means with the same letters are not significantly different according to Tukey's HSD test (α=0.05).

3.4.2 Interaction of crop rotation and herbicide strategy

The continuous use of high resistance risk herbicides (HS4), finally resulted in highest blackgrass densities in all CR at both locations (Figure 4). Densities in HS4 were significantly higher compared to HS2 and HS3 already in the third or rather fourth year. Concurrently, the continuous use led to a loss in efficacy of HRAC A and B herbicides. The resistance development was faster and started earlier at Wurmberg, where a reduced control efficacy of herbicides was already observed prior to the experiment (Figure 6). Highest blackgrass densities (2086 heads m^{-2}) were counted in CR1 in year 4 at Wurmberg. After five years blackgrass heads m^{-2} amounted to 917 in CR1, 770 in CR2 and 366 in CR3 at Ihinger Hof and 1850, 1138 and 477 at Wurmberg, respectively.

HS3 with a medium selection pressure failed in winter wheat in the third year due to an omitted post-emergence herbicide application and a late emergence of blackgrass approximately 4 weeks after application date. In contrast to HS4, blackgrass densities further decreased in HS3 in the fourth and fifth year. Results have shown that CRs including spring crops and herbicide MOA rotations (HS2) were capable to suppress blackgrass over the five-year study. An exception was year 2, when the AIs isoproturon and tembotrione were used.

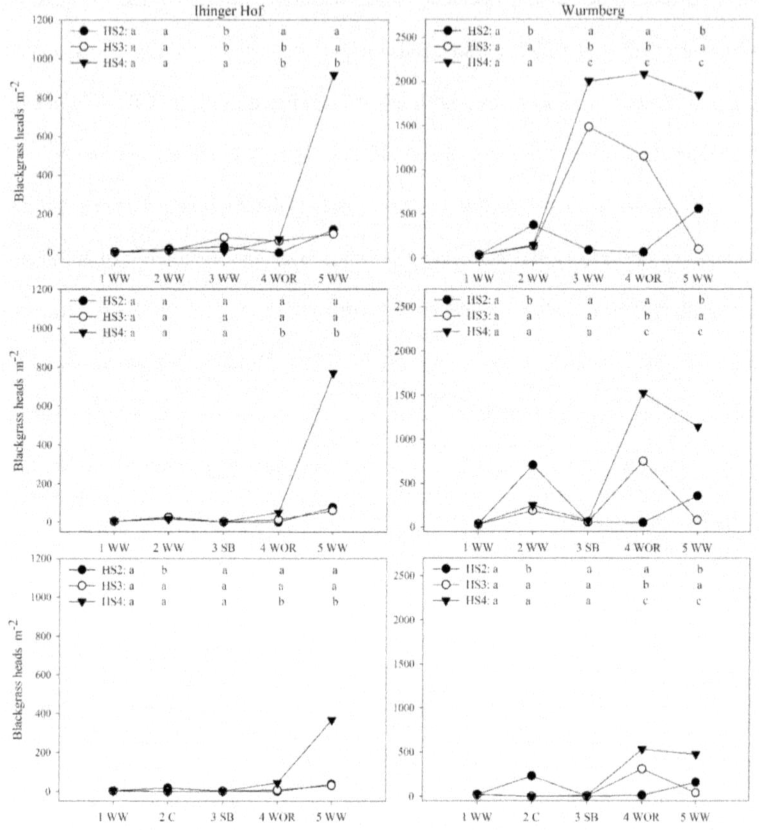

Figure 4. Blackgrass heads m^{-2} in the herbicide strategies (HS) 2 – 4 and the three crop rotations (CR) at Ihinger Hof and Wurmberg. Means with the same letters are not significantly different according to Tukey's HSD test (α=0.05).

3.4.3 Effect of blackgrass infestation on crop yield

No significant crop yield loss was determined in winter wheat (CR1) at Ihinger Hof in the first and second year. At Wurmberg, the higher infestation already resulted in a significant winter wheat yield loss in the second year. Corn yield and the respective yield parameters, corn grain (Ihinger Hof) and corn silage (Wurmberg) were significantly reduced in HS1. In year 3, high yield losses were measured in HS1 when winter wheat was cultivated for the third time (CR1) at both locations. In spring barley (CR2 & 3) no yield losses were determined. In year 4 and 5, crop yields in HS1 were significantly lower compared to the other HSs. In year 5, also the yields in HS4 were negatively affected (Figure 5).

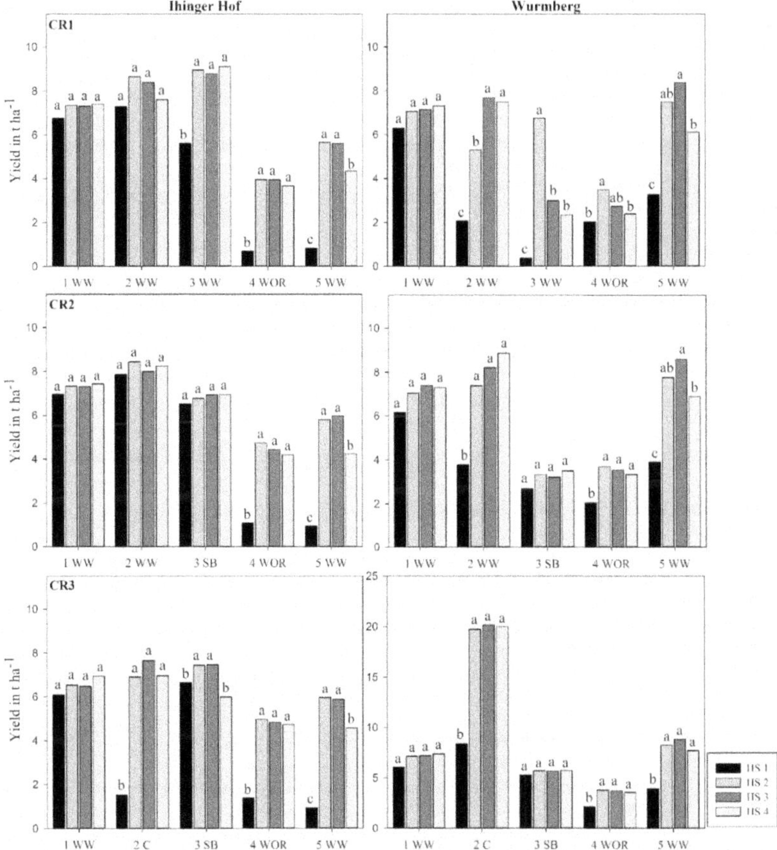

Figure 5. Crop yield in the three crop rotations (CR) and four herbicide strategies (HS) at both locations. Means with the same letters are not significantly different according to Tukey's HSD test (α=0.05).

3.4.4 Herbicide resistance investigations

3.4.4.1 Greenhouse bioassay

In Figure 6 the efficacies of the AIs, pinoxaden (DEN), meso- and iodosulfuron (SU), pyroxsulam (TP) and isoproturon (U) are shown and compared with the efficacies before the experiment started (black bars). At Wurmberg the herbicide efficacy of meso- and iodosulfuron (94 %), pyroxsulam (85 %) and isoproturon (87 %) was lower compared to Ihinger Hof, where herbicide efficacies close to 100% were detected. Therefore, regarding pyroxsulam and isoproturon the blackgrass population at Wurmberg was classified as 1*R? according to the R-Scaling by Moss *et al.* (1999), assuming that some selection has already occurred at Wurmberg before the experiment started.

At Ihinger Hof the populations sampled in HS1 and HS2 remained susceptible (S) and have shown herbicide efficacies of ≥ 95 % with regard to all tested AIs. For pinoxaden HS3 was rated R? with efficacies of 91 (CR2) and 81% (CR1). HS4 has shown the lowest herbicide efficacies and a confirmed resistance (RR) could be determined for pinoxaden and pyroxsulam in CR1, with efficacies of 71 % and 73 %, respectively.

At Wurmberg HS4 and HS3 were most affected. Confirmed resistances were detected for pinoxaden, meso- and iodosulfuron as well as pyroxsulam. The rating scaled between 36 % (RRR) and 80 % (R?) in CR1. Regarding pinoxaden, even in HS2 and HS1 where less selection pressure was exerted it was shown that populations shifted to R?. However, herbicide efficacy even increased if no selection pressure was exerted as shown for pyroxsulam. Generally, it can be stated that the more winter annual crops were cultivated in a CR, the lower control efficacies were observed in the in HS.

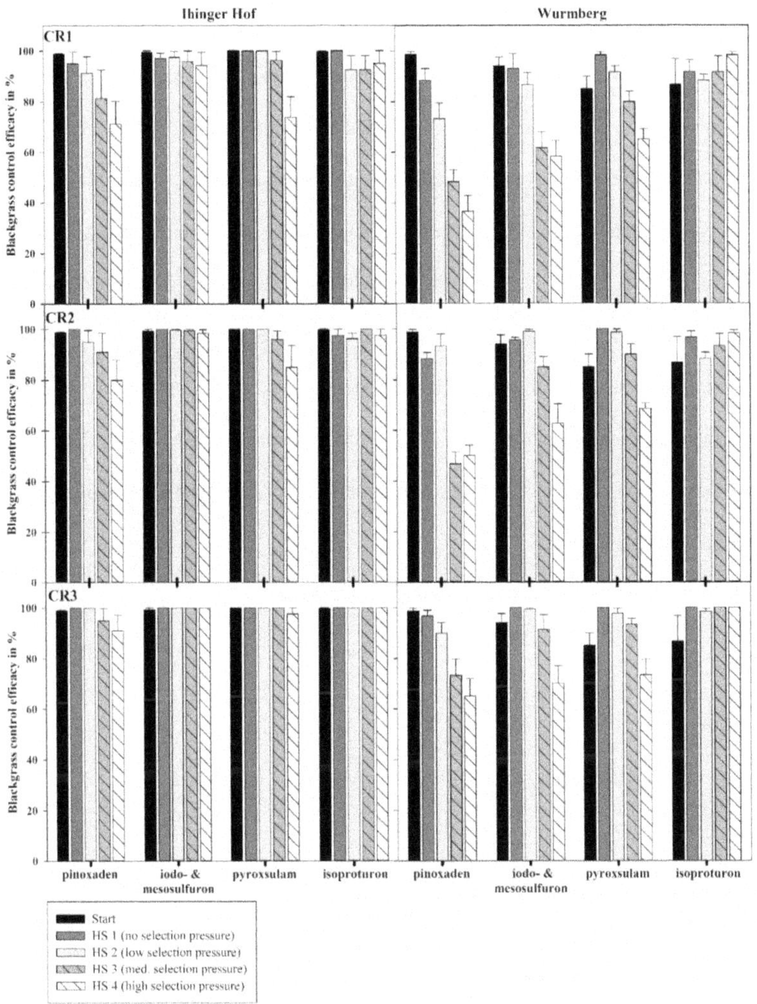

Figure 6. Herbicide control efficacies on blackgrass originating from crop rotation (CR) and herbicide strategy (HS) plots of the two experimental locations as investigated in a greenhouse bioassay. Bars represent the standard error of the mean.

3.4.4.2 Investigation of target-site-resistance (TSR)

A mutation in the ACCase gene was found in 52 % of the tested plants originating from Ihinger Hof, which survived the pinoxaden application (Table 7). For plants which survived the treatment with ALS – inhibitors, no mutations could be determined in the ALS gene. In plant material from Wurmberg, mutations in the specific gene were found in 20 % and 26 %, respectively. For PSII – inhibitors no mutations were found in plants from both locations. For plants which survived the herbicide treatment but no mutation was found within the relevant gene loci, other resistance mechanisms must be responsible. Nevertheless, plants originating form HS4, CR1 or a combination have shown multiple mutations and mutations which did not concern the specific gene loci of the applied herbicides at both locations.

Table 7. TSR – analysis of the surviving plants in the bioassay.

HRAC	No. tested	Positive	Negative	Match	Mutation in the target site
			Ihinger Hof		
A	23	13 57 %	10 43 %	12 52 %	5 x 1781 (Ile/Leu) 2 x 2027 (Trp/Cys) 2 x 2041 (Ile/Asn) 1x 1781 (Ile/Leu) + 2027 (Trp/Cys) 1x 197 (Pro/His)* 1x 1781 (Ile/Leu) + 197 (Pro/His)* 1x 2027 (Trp/Cys) + 2041 (Ile/Asn) + 197 (Pro/His)*
B	18	1 6 %	17 94 %	0 0 %	1 x 178 (Ile/Val)*
			Wurmberg		
A	46	10 22 %	36 78 %	9 20 %	9 x 1781 (Ile/Leu) 1 x 197 (Pro/His)*
B	58	19 33 %	39 67 %	15 26 %	10 x 197 (Pro/His) 4 x 574 (Trp/Leu) 1x 197 (Pro/His) + 1781 (Ile/Leu)* 4 x 1781 (Ile/Leu)*

No. Tested = total number of tested samples. Positive = number of tested samples with mutation. Negative = number of tested samples without mutation. Match = mutation was found in the gene loci which is relevant for the applied herbicide . * mutation does not explain the survival as it affects a gene loci which is not affected by the applied herbicide.

3.5 Discussion

Although the locations were quite different with regard to its initial situation and the different starting dates, similar results were obtained. Blackgrass densities increased rapidly when no blackgrass control was conducted. A fast resistance development was recognized when the selection pressure exerted by herbicides was high. Resistance development was higher the more winter annual crops were cultivated. It was shown that summer annual crops reduced infestations and resistance development. Exclusively the rotation of herbicide MOA in combination with summer annual crops was able to prevent resistance development and reduce already established resistances. Therefore, our assumptions form the hypotheses were confirmed. Our results stay in line with investigations of Moss (1987a), who stated that blackgrass increased by up to 10-fold per year when the population remained uncontrolled. Moreover, Chauvel et al. (2001) as well as Lutman et al (2013) demonstrated a reduction of infestations due to the cultivation of spring barley. Additionally, the present results confirmed the findings of Powles and Yu (2010) who showed that a rapid shift of populations towards resistant biotypes can occur by herbicide selection.

Blackgrass has a high preference to winter cereals (Naylor, 1972a; Blair et al., 1999). It was propagated faster in winter cereals than in winter oilseed rape. A 1.5 – fold increase was observed in winter oilseed rape compared to 3.9-5.7 – fold increase per year in winter cereals. Colbach et al. (2007) also demonstrated lower infestations of blackgrass when winter oilseed rape was cultivated. We assume, that due to the early sowing dates of winter oilseed rape in the first week of September a higher proportion of blackgrass seeds were still dormant compared to sowing dates of winter cereals in October. Winter oilseed rape is more competitive against blackgrass than winter cereals developing a dense canopy with higher leaf area index than winter wheat (Christen and Fried, 2007). Nevertheless, spring crops are most effective to control blackgrass with a 3.5-22 – fold decrease. A reduced blackgrass density leads to a reduced weed seed entry into the soil and therefore a lower infestation in the following years. This effect was already estimated in the studies of Colbach et al. (2006; 2010) using the ALOMYSYS model. The effect is higher the more summer annual crops are cultivated in the rotation. Beside the lower infestations, spring crops like sugar beets, potatoes and corn enable mechanical weed control measures such as hoeing. Additionally, wider crop rotations allow the use of a greater diversity of herbicide MOAs or at least chemical families which is advantageous from a resistance management perspective. For example, herbicides such as cyclohexadiene (HRAC – A), metribuzin (C1), aclonifen (F3) or propyzamid (K1) can be used in broader crop rotations as they are not registered for the use in cereals. The present study shows that the use of herbicides with a lower efficacy can be reasonable from a resistance management perspective, if this allows a rotation of MOA and a longer gap between herbicides with the same MOA.

Herbicide resistance is defined as a consequence of naturally occurring changes in plants leading to the ability to survive herbicide applications, which are lethal for the wild type (Powles and Preston, 1995). Our results and also investigations of Powles and Yu (2010) have demonstrated that an intensive use of herbicides with the same MOA resulted in a fast selection of resistant biotypes. Mortimer et al. (1992) concluded that there is an increasing risk of resistance development for species like blackgrass which are characterized by cross-pollination, high growth rates and non-persistent seeds combined with high selection pressure. Foster et al. (1993) stated, that compared to other weed control strategies, the use of herbicides exerts the highest selection pressure with control efficacies by up to 99.99 %. This became obvious in the bioassay results. The higher blackgrass densities and selection pressure were, the faster the resistance development was and the herbicide efficacies were observed.

By molecular genetic analysis of the target sites, the resistance mechanism of averaged 25 % plants could be clarified. For the rest of the plants other resistance mechanisms like an overexpression in the target-site, changes in the promotor region of the gene or a NTSR might be responsible for the absent herbicide efficacy (Délye et al., 2013; Heap, 2014). TSR analysis has shown that only for plants originating from variants with very high blackgrass densities (CR1 and HS4) multiple mutations within one plant were found. Our assumption, which was also discussed in Hicks et al. (2018), is that the higher infestations result in a higher occurrence of genetically modified biotypes. This leads to an increased selection risk, especially when herbicides with the same MOAs are used.

3.6 Conclusion

The recent study illustrates the need for IWM strategies consisting of multiple tactics to control blackgrass effectively. Wider crop rotations containing winter and spring crops reduce blackgrass densities, prevent resistance development and facilitate a rotation of herbicides with different MOAs. Only a combination of spring- and autumn sown crops with a rotation of herbicide MOA sufficiently suppressed blackgrass at those two locations. Densities of resistant blackgrass were higher the more frequent ALS- and ACCase-inhibitors were applied in the rotation. Therefore, the results of this study support the legislative regulations for some herbicides to use them only once in a 4-years' rotation.

3.7 Acknowledgements

The authors wish to thank Mr. Capezzone from the Biostatistics Department at the University of Hohenheim for his assistance regarding data analysis. Furthermore, we like to thank the colleagues at Ihinger Hof Research Station, Hartmut Weeber and all technical assistants and students for their support to conduct the experiments and collect the data. This study was funded by the German Federal Office for Agriculture and Food.

Chapter IV

–

A long-term study of crop rotations, herbicide strategies and tillage practices: Effects on *Alopecurus myosuroides* Huds. abundance and contribution margins of the cropping systems

Alexander K. Zeller[a], Yasmin I. Kaiser[b] and Roland Gerhards[a]

[a]*University of Hohenheim, Institute of Phytomedicine, Department of Weed-Science, Otto-Sander-Str. 5, 70599 Stuttgart, Germany*

[b]*FMC Agricultural Solutions, Westhafenplatz 1, 60327 Frankfurt am Main, Germany; previously University of Hohenheim*

Submitted to: *Crop Protection*

4 A long-term study of crop rotations, herbicide strategies and tillage practices: Effects on Alopecurus myosuroides Huds. abundance and contribution margins of the cropping systems

4.1 Abstract

Alopecurus myosuroides Huds. (blackgrass) is an abundant weed in many European countries causing high yield losses in winter-cereals and increasing costs for weed control. Several agronomic measures to control blackgrass are available, however long-term studies with several factors affecting blackgrass are rare. In this seven-year study, three crop rotations with different proportions of winter-annual crops, four herbicide strategies with various selection pressure and four different tillage measures were combined and their interaction was tested. Aim was to investigate the effects on blackgrass abundance, crop yields and contribution margins. False seedbed preparation using a harrow (shallow) and a rotary harrow (deep) reduced blackgrass infestations by up to 70 % compared to solely reduced conservation tillage. In combination with crop rotations including spring crops and alternating herbicide mode of actions, control efficacies of up to 100 % were achieved. The predominant cultivation of winter-annual crops combined with conservation tillage and the continuous use of the same active ingredients increased blackgrass densities more than 10-fold compared to the initial infestation rate. Further, contribution margins tend to be highest in the cropping systems that combine all three weed control measures. These results clearly highlight the benefits of integrated weed management strategies in regard to weed control and economic sustainability.

Keywords: blackgrass control, harrow, false seedbed, preventive measures, integrated weed management

4.2 Introduction

Blackgrass, *Alopecurus myosuroides* Huds. (Poales: Poaceae), is one of the most abundant weeds of autumn-sown crops in Europe, mainly in the western and central parts (Moss et al., 2007). It is a winter-annual grass weed, which predominantly germinates from September to October and is propagated by seeds (Naylor, 1972a; Moss, 1985). Blackgrass seeds have a short dormancy cycle and thus, a high germination rate (Naylor, 1972b; Froud-Williams, 1985). Cussans et al. (1996) demonstrated that 90 % of blackgrass seeds germinated from soil layers not deeper than 0.05 m, as light is required to initiate the germination process (Colbach et al., 2002a; Anderson and Espeby, 2008). If burial is deeper than 0.05m, the majority of seeds exhibit secondary dormancy (Cussans et al., 1996). Soil seed bank of blackgrass is reduced by up to 70-80 % per year if seeds are buried (Cousens and Moss, 1990). Therefore, simplified crop rotations with high proportions of autumn-sown crops, early sowing dates of winter cereals and reduced tillage practices promoted blackgrass infestations (Gerhards et al., 2013). Due to the continuous use of herbicides with the same mode of action (MOA), many herbicide-resistant populations have developed in European countries (Moss et al., 2007; Moss, 2017; Heap, 2019). If blackgrass populations cannot be controlled, blackgrass abundance increases rapidly and can case high yield losses (Moss, 1987b; Blair et al., 1999).

To control blackgrass populations and prevent resistance development multiple integrated weed management (IWM) strategies are needed (Lutman et al., 2013; Moss, 2017). Those IWM strategies include preventive measures to suppress weeds and increase crop competitiveness as well as direct measures, if the economic weed control threshold is still exceeded (Swanton and Weise, 1991; Mortensen et al., 1995; Buhler, 2002). So far, spring cropping and competitive crop cultivars (Chauvel et al., 2001; Lutman et al.,

2013), inversion tillage (Pollard et al., 1982; Moss, 1987a; Turley et al., 1996) later sowing dates in autumn (Melander, 1995; Hurle, 1993) as well as cover cropping (Brust et al., 2014) have been identified to suppress blackgrass. However, long-term field studies examining different IWM strategies and analyzing the interactive impact on blackgrass abundance as well as crop yield are rare.

The aim of this study was to quantify effects of crop rotation (CR), herbicide strategies (HS) and tillage measures (T) on blackgrass abundance as well as its impact on crop yield and contribution margins (CM). The hypotheses were that a) including spring crops into the CR reduces blackgrass infestations; that b) false seedbed preparation decreases blackgrass abundance compared to conservation tillage practices, and that c) the CMs of cropping systems combining different IWM strategies are higher than weeding system with high reliance on herbicides.

4.3 Materials and methods

4.3.1 Field experiments

Field experiments started in autumn 2011 at Ihinger Hof (Renningen; 48.74° N, 8.92° E, 478 m ASL) and 2012 at Wurmberg (48.86° N, 8.83 °E, 450 m ASL). Average annual temperatures at these locations are 9.1 °C and 10.6 °C with a precipitation amount of 825 mm and 736 mm, respectively. The soil texture is a clayey loam at both sites. Experiments were set up with a randomized split-plot design with four replicates.

The main plot factor was the CR with a plot size of 24 x 24 m. The CRs differed in their proportions of winter annual crops but all had a duration of four years. CR1 (100 % winter-annual crops) included three years of winter wheat followed by winter oil-seed rape. In CR2 (75 % winter-annual crops), spring barely was grown in the third year instead of winter wheat. In CR3 (50 % winter annual crops), corn was grown in the second year and spring barely in the third year.

The sub-plot factor was HS with a sub-plot size of 6 x 12 m. In HS1, the untreated control, no herbicides against blackgrass were applied and no selection pressure on blackgrass was exerted. Selection pressure in HS2 was low, because MOAs were changed in every year and each MOA was used only once during a CR period. In HS3, a medium selection pressure was exerted by applying herbicide-mixtures or -sequences and in HS4, a high selection pressure was induced using herbicides with the same MOAs in every year. Dicotyledonous weeds were controlled in all HSs by herbicides, which had no efficacy on blackgrass. All utilized herbicides were applied at the recommended rates (Table 8). Application was carried out with a self-propelled plot sprayer (Schachtner-Gerätetechnik, Ludwigsburg, Germany), which was calibrated for a volume of 200 L ha^{-1} and a spray pressure of 170 kPa. Air injector double flat fan nozzles (02-120° IDTK, Lechler, Metzingen, Germany) were used for application. The 192 sub-plots were separated by 0.4 m strips, that were treated with glyphosate (1440 g AI ha^{-1}, Clinic®, Nufarm GmbH, Cologne, Germany) to reduce cross-pollination.

Table 8. Utilized herbicides of the herbicide strategies (HS) in the different crop rotations (CR) from the year one to seven with time of application (timing), tradename and formulation, active ingredient (AI), herbicide MOA and application rate.

Year	CR / Crop	Timing	HS	Tradename / formulation	Active ingredient	HRAC-Code	Rate (g AI ha^{-1})
1	1-3 WW	spring	2-4	Broadway™ WG[1]	pyroxsulam	B	15
2	1+2 WW	spring	2	Arelon® SC[2]	isoproturon	C1	1500
		spring	3	Atlantis® WG[3]	mesosulfuron + iodosulfuron	B	9 + 1.8
		spring	4	Broadway™ WG[1]	pyroxsulam	B	15
	3 C	spring	2	Laudis® OD[3]	tembotrione	F2	88
		spring	3	Kelvin® OD[4]	nicosulfuron	B	32
		spring	4	Elumis® OD[5]	nicosulfuron	B	37.5
3	1 WW	autumn,	2	Herold® SC[6];	flufenacet + diflufenican;	K3 + F1;	240 + 120;
		spring		Traxos® EC[5]	clodinafop + pinoxaden	A	30 + 30
		autumn	3	Herold® SC[6]	flufenacet + diflufenican	K3 + F1	240 + 120
		spring	4	Broadway™ WG[1]	pyroxsulam	B	15
	2+3 SB	spring	2-4	Axial® 50 EC[5]	pinoxaden	A	60
4	1-3 WOR	autumn,	2	Butisan® Aqua EC[4];	metazachlor + pendimethalin;	K3 + K1;	400 + 341.25;
		winter		Kerb™ Flo SC[1]	propyzamid	K1	750
		autumn	3	Butisan® Aqua EC[4];	metazachlor + pendimethalin;	K3 + K1;	400 + 341.25;
				Select® 240 EC[2]	clethodim	A	121
		autumn	4	Gallant® Super EC[1]	haloxyfop	A	52
5	1-3 WW	autumn	2	Boxer® EC[5]	prosulfocarb	N	4000
		autumn	3	Lexus® EC[2] + Malibu® EC[4]	flupyrsulfuron; pendimethalin + flufenacet	B; K1 + K3	10; 1200 + 240
		spring	4	Broadway™ WG[1]	pyroxsulam	B	15
6	1+2 WW	spring	2	Atlantis® WG[3]	mesosulfuron + iodosulfuron	B	9 + 1.8
		autumn,	3	Herold® SC[6];	flufenacet + diflufenican;	K3 + F1;	240 + 120;
		spring		Atlantis® WG[3]	mesosulfuron + iodosulfuron	B	9 + 1.8
		spring	4	Broadway™ WG[1]	pyroxsulam	B	15
	3 C	spring	2-4	Elumis® OD[4]	tembotrione	F2	37.5
7	1 WW	autumn,	2	Herold® SC[6];	flufenacet + diflufenican;	K3 + F1;	240 + 120;
		spring		Lentipur SC[7]	chlorotoluron	C2	2100
		autumn	3	Herold® SC[6];	flufenacet + diflufenican;	K3 + F1;	240 + 120; 9
				Atlantis® WG[3]	mesosulfuron + iodosulfuron	B	+ 1.8
		spring	4	Broadway™ WG[1]	pyroxsulam	B	15
	2-3 SB	spring	2-4	Axial® 50 EC[5]	pinoxaden	A	60

CR = crop rotation, HS = herbicide strategy, AI = active ingredient, WW = winter wheat, C = corn, SB = spring barley, WOR = winter oilseed rape; SC = suspension concentrate, WG = water dispersible granule, OD = oil dispersion, EC = emulsion concentrate; [1]Corteva Agriscience™ (Munich, Germany), [2]FMC Agricultural Solutions (Stade, Germany), [3]Bayer CropScience GmbH (Langenfeld, Germany), [4]BASF (Ludwigshafen, Germany), [5]Syngenta Agro GmbH (Maintal, Germany), [6]Adama GmbH (Cologne, Germany), [7]Lotus Agrar GmbH (Mannheim, Germany).

Until year five, only conservation tillage operations were performed not deeper than 0.12 m. The tillage consisted of removing stubble (Dyna-Drive Pro, Bomford Turner Ltd, Worcestershire, UK) followed by one chisel plough treatment (K300, Kerner GmbH, Aislingen, Germany) as well as seedbed preparation with a rotary harrow (Zirkon 8/300, Lemken GmbH, Alpen, Germany) and seeding. After harvest of winter wheat in year five in 2016 (Ihinger Hof) and 2017 (Wurmberg) four different T measures were included as factor into the field trial. The HS sub-plots were divided into further sub-plots with a plot size of 6 x 3 m. The different T measures were distributed over the entire experimental site using a Latin Square design. T1 represented conservation tillage as practiced in the previous five years. In T2, inversion tillage at a depth of 0.25 m was performed with a mouldboard plough (EurOpal 5 3 N 90, Lemken GmbH Alpen, Germany) followed by seedbed preparation and seeding like in T1. In year seven, the plots were not ploughed and T2 tillage measures corresponded to T1. In T3 and T4 a false seedbed preparation was added to the reduced tillage. In T3 this was performed with a harrow at a depth of 0.05 m (Aerostar, Einböck GmbH, Dorf an der Pram, Austria) on two treatment dates in mid and end of September. The false seedbed preparation in T4 was carried out with a rotary harrow at 0.1 m depth once in mid of October. Approximately 14 days after false seedbed preparation same seeding was carried out like in T1 (Figure 7). For cereals a single disc drill (750 A, John Deere, Walldorf, Germany) with a row distance of 0.12 m and for corn a precision air seeder (Maxima 2, Kuhn GmbH, Schopsdorf, Germany) with a row distance of 0.75 m were used.

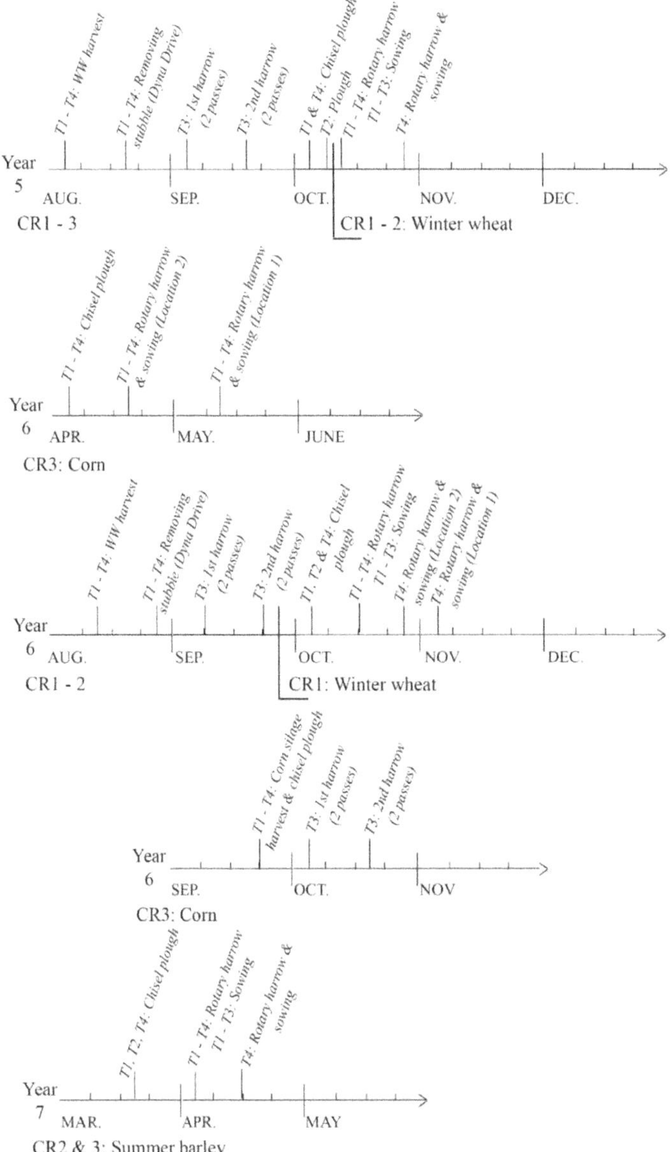

Figure 7. Timelines of the four tillage measures (T) in the three crop rotations (CR). Months are separated in 4 weeks and lines define the approximate time of the treatment. If not noted, timing of operations was applied at both locations at the same dates.

The environmental conditions in the experimental years 2016 – 2019 were atypical at both locations, especially in year 2018. Compared to the long-time average, temperatures were increased Ø +1.6°C at Ihinger Hof and Ø +2.3°C at Wurmberg as well as precipitation was decreased ∑ -215 mm and ∑ -219 mm, respectively (Figure 8). The lower precipitation and the high temperatures in 2018 affected the annual blackgrass infestation, crop yields and sales revenues.

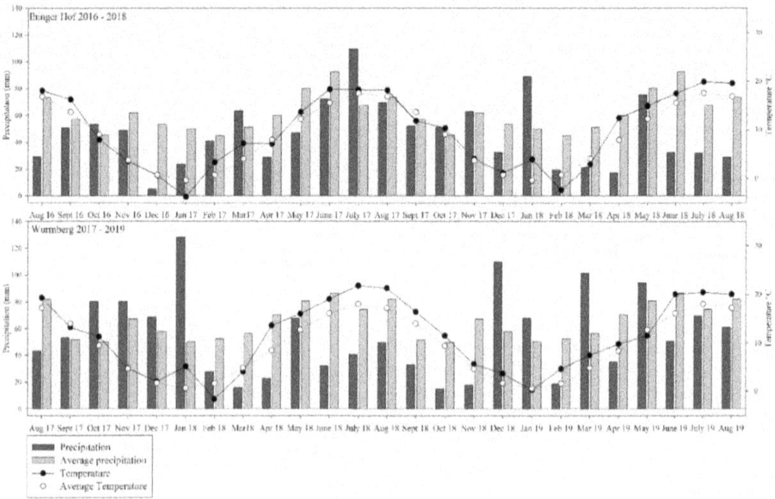

Figure 8. Temperature in °C and precipitation in mm in the experimental years 2016 – 2018 at Ihinger Hof and 2017 – 2019 at Wurmberg, compared to the long-time average of the last 30 years. The weather data are based on weather stations at or rather near the experimental sites (LTZ, 2019).

4.3.2 Data collection

Blackgrass abundance was assessed by counting heads m^{-2} in 5 randomly placed frames (0.4 m^2) per plot at flowering stage of the crop. Crop yields were recorded in the core (2 x 5 m) of each sub-plot using a plot combine harvester for cereals (Zürn170, Zürn

Harvesting GmbH & Co. KG, Schöntal-Westernhausen, Germany) or a plot field chopper for corn (MH500, Deutz Fahr GmbH, Lauingen, Germany).

4.3.3 Data analysis

4.3.3.1 Calculation of contribution margins

CMs represent the sales revenues minus the variable costs. CMs were calculated for each CR x HS x T with an Excel (Microsoft® Office, Santa Rosa, California, USA) database. Machinery costs for tillage and spraying operations were obtained from the KTBL database (KTBL, 2019) of the German Federal Ministry of Food and Agriculture. Prices for herbicides and seeds were collected from the providers (Agrolandis, 2015; Schweiger, 2018; Agraronline, 2019; BayWa, 2019). Fertilizers were calculated with 1.19 € kg^{-1} N, 0.95 € kg^{-1} P_2O_5, 0.71 € kg^{-1} K_2O and 0.60 € kg^{-1} MgO (LEL, 2019). Sale revenues for crops were based on average prices from 2017 – 2019 (LEL, 2019), which were 172.7 € t^{-1} for winter wheat, 199.3 € t^{-1} for spring barley and 37.9 € t^{-1} corn silage multiplied by the actual obtained yields in the respective years. Price volatility, costs for fungicides and insecticides as well as agricultural direct- and premium-payments were not taken into account.

4.3.3.2 Statistical analysis

Data was analyzed with the statistical software SAS® (V9.4, SAS Institute Inc, NC, USA). A linear mixed effect model was used to evaluate the effects of CR, HS, T and year on blackgrass abundance and crop yield. A dummy variable was used to include the factor tillage into evaluation starting in year six. Data of crop yields were log transformed and blackgrass abundance were square root transformed to homogenize variances and to normalize distribution. Means were compared using the Tukey's HSD test at a significance level of $\alpha \leq 0.05$ according to the SAS macro described by Piepho (2004; 2012). Both sites were analyzed separately because of the different blackgrass infestation and resistance levels as well as the different starting times of the experiment. Back transformed means are presented in tables and figures. Figures were created with Sigmaplot V12.5 (Systat software GmbH, Erkrath, Germany) and Adobe Photoshop CS2 (Adobe Inc., California, USA).

4.4 Results

4.4.1 Field experiment

4.4.1.1 Effects on blackgrass infestation

The interaction of the factors CR, HS and year were highly significant in the first five years of the experiment at both locations. After the factor T was added to the experiment in year six, the significance of the interaction continued (Table 9).

Table 9. Level of significance for the factors crop rotation (CR), herbicide strategy (HS), tillage (T) and year in the experimental years 1-5 and 1-7 of both locations.

Year	Effect	Level of significance (p-value)
1 – 5	CR x HS x year	< 0.0001
1 – 7	CR x HS x T x year	< 0.0001

The experiment started with 14 blackgrass heads m^{-2} at Ihinger Hof and 201 heads m^{-2} at Wurmberg. Abundance increased to 5347 heads m^{-2} at Ihinger Hof and 3424 heads m^{-2} at Wurmberg after five years, if no chemical weed control was applied (HS1) and only winter annual crops were grown (CR1). Due to the cultivation of spring crops, blackgrass abundance was 33 % lower in CR2 and 50 % lower in CR3 at Ihinger Hof compared to CR1 and reduced by 7 % and 32 % at Wurmberg, respectively. Blackgrass densities were significantly reduced in the third year at both locations due to the cultivation of spring barley instead of winter wheat. At the latest in year four, herbicide efficacies started to decrease at both locations in HS4, where the same MOAs were used every year. In year five, infestations amounted to 400 (CR3) - 2000 (CR1) heads m^{-2}, due to herbicide resistance development. The other herbicide strategies (HS2 & 3) controlled the blackgrass infestations in the field sufficiently, especially in combination with spring cropping. However, resistance investigations by greenhouse bioassay and molecular genetic analysis of the target-site confirmed a resistance development in HS3 and HS4 (data not shown). Only the combination of 50 % winter annual crops (CR3) and the rotation of herbicide MOAs in every year (HS2) prevented an increase of blackgrass abundance and herbicide resistance development (Zeller et al. 2019, submitted). Data analysis after year five showed significant increases in blackgrass abundance in particular CRxHS combinations, therefore from year six onwards the additional factor of T was tested to reduce blackgrass.

Blackgrass abundance as affected by CR, HS, T and year are shown in Table 10 for both locations. As in the previous years, blackgrass abundance was significantly decreased due to the cultivation of spring crops (CR2 & 3). However, only the combination of spring cropping and herbicides suppressed blackgrass infestations sufficiently. Independent of the CR, highest blackgrass densities were counted in HS4, which used the same MOAs in every year, and T1, the treatment with solely reduced tillage. Inversion tillage (T2) and the false seedbed preparation (T3 & 4) significantly decreased blackgrass abundance in comparison to T1. False seedbed preparation was in majority significantly better than inversion tillage, especially in year seven. With increasing blackgrass infestation levels, the effect of the tillage measures (T2-T4) increased.

Table 10. Amount of blackgrass heads m^{-2} in the three crop rotations (CR), four herbicide strategies (HS) and four tillage (T) measures in years 6 and 7 at the experimental locations Ihinger Hof and Wurmberg.

Ihinger Hof																	
Blackgrass heads m^{-2}		Year 6 (2017)								Year 7 (2018)							
		HS1		HS2		HS3		HS4		HS1		HS2		HS3		HS4	
CR1	T1	3098	a	34	a	7	a	1128	a	1225	a	167	a	252	a	489	a
	T2	1899	c	28	a	2	a	441	b	1113	a	109	ab	148	b	302	b
	T3	2373	b	23	a	6	a	518	b	602	b	59	b	91	b	163	c
	T4	1413	d	12	a	2	a	190	c	373	c	16	c	0	c	44	d
CR2	T1	2263	a	27	a	6	a	658	a	1103	a	53	a	47	a	369	a
	T2	1428	b	16	a	3	a	453	b	892	a	16	ab	10	bc	178	b
	T3	2222	a	22	a	5	a	474	b	580	b	12	bc	14	ab	114	b
	T4	623	c	20	a	1	a	109	c	67	c	0	b	0	c	8	c
CR3	T1	73	a	14	a	4	a	13	a	667	a	8	a	2	a	220	a
	T2	34	a	0	b	0	a	6	a	431	b	2	a	2	a	80	b
	T3	44	a	5	ab	2	a	3	a	284	c	0	a	0	a	69	b
	T4	63	a	6	ab	1	a	6	a	34	d	0	a	0	a	6	c
Wurmberg																	
Blackgrass heads m^{-2}		Year 6 (2018)								Year 7 (2019)							
		HS1		HS2		HS3		HS4		HS1		HS2		HS3		HS4	
CR1	T1	2872	a	1381	a	119	a	2100	a	3563	a	853	a	145	a	1960	a
	T2	2052	b	734	b	34	ab	1213	b	3377	a	584	ab	148	a	1896	ab
	T3	2203	b	844	b	53	ab	1331	b	3263	a	449	b	143	a	1666	ab
	T4	1150	c	372	c	28	b	850	c	2976	a	454	b	111	a	1504	b
CR2	T1	2338	a	1034	a	150	a	1631	a	452	a	128	a	142	a	166	a
	T2	1353	bc	650	bc	81	a	1213	bc	344	ab	84	ab	50	ab	110	a
	T3	1680	b	928	ab	125	a	1391	ab	319	ab	25	b	33	b	60	a
	T4	1166	c	416	c	50	a	969	c	223	b	49	b	38	b	62	a
CR3	T1	183	a	0	a	0	a	0	a	45	a	3	a	6	a	12	a
	T2	109	a	0	a	0	a	0	a	23	a	4	a	6	a	6	a
	T3	144	a	0	a	0	a	0	a	11	a	1	a	5	a	4	a
	T4	152	a	0	a	0	a	0	a	13	a	0	a	3	a	6	a

Means within one HS and CR followed by the same letter are not significantly different according to Tukey's HSD test ($\alpha=0.05$)

4.4.1.2 Effects on crop yield

Crop yields as affected by CR, HS, T and year are shown in Table 11 for both locations. Year was the biggest determinant on crop yield. The high temperatures and the low precipitation in 2018 resulted in lower crop yields in all treatments, especially at Ihinger Hof. Yields of 2017 and 2019 are more similar to the previous years one to five (Zeller et al., 2019, submitted). If no herbicides against blackgrass were applied (HS1) yields were lower compared to the other HSs and yields of HS4 continued decreasing, if winter annual crops were cultivated exclusively (CR1). Due to the cultivation of spring crops (CR2 & 3) high yields were achieved, even when no herbicides were applied. Particularly spring barley achieved high yields. Nonetheless, the various tillage measures (T1-4) resulted in significantly different blackgrass densities, crop yields did in majority not differ significantly (Table 11).

Table 11. Yield in t ha^{-1} in the three crop rotations (CR), four herbicide strategies (HS) and four tillage measures (T) in years six and seven at the two experimental locations Ihinger Hof and Wurmberg.

Yield in t ha^{-1}		Ihinger Hof															
		Year 6 (2017)								Year 7 (2018)							
		HS1		HS2		HS3		HS4		HS1		HS2		HS3		HS4	
CR1	T1	4.2	b	7.0	a	6.9	a	7.1	a	3.6	a	4.7	a	4.2	a	4.7	a
	T2	5.0	b	7.4	a	7.3	a	7.3	a	3.7	a	4.9	a	4.3	a	4.5	a
	T3	5.0	b	7.4	a	7.6	a	7.6	a	4.0	a	5.5	a	4.9	a	4.6	a
	T4	6.2	a	7.3	a	7.5	a	7.7	a	3.7	a	3.0	b	3.3	b	4.4	a
CR2	T1	4.3	b	7.4	a	8.1	a	7.2	a	2.0	b	2.7	bc	2.6	b	2.6	b
	T2	5.0	b	7.3	a	8.0	a	7.5	a	1.9	b	2.9	b	2.8	b	2.4	b
	T3	4.6	b	7.5	a	7.6	a	7.4	a	2.7	a	3.7	a	3.6	a	3.8	a
	T4	7.3	a	7.8	a	7.9	a	8.1	a	2.6	a	2.3	c	2.4	b	2.4	b
CR3	T1	31.3	a	33.0	a	32.8	a	32.9	a	2.6	b	3.7	a	3.3	a	3.4	a
	T2	28.8	a	34.3	a	32.4	a	33.3	a	2.7	b	3.9	a	3.5	a	4	a
	T3	31.9	a	33.3	a	31.4	a	32.9	a	3.7	a	4.1	a	3.7	a	4.1	a
	T4	29.0	a	31.0	a	33.0	a	31.0	a	2.0	c	2.0	b	1.9	b	1.8	b
Yield in t ha^{-1}		Wurmberg															
		Year 6 (2018)								Year 7 (2019)							
		HS1		HS2		HS3		HS4		HS1		HS2		HS3		HS4	
CR1	T1	3.6	a	5.1	a	6.3	a	5.7	a	3.5	a	5.8	a	7.6	a	4.6	a
	T2	3.1	a	5.5	a	7.2	a	5.6	a	3.7	a	6.1	a	8.4	a	4.6	a
	T3	3.0	a	5.0	a	7.3	a	4.9	a	3.4	a	6.6	a	7.4	a	4.7	a
	T4	3.5	a	5.9	a	7.2	a	5.3	a	3.8	a	7.3	a	7.2	a	4.9	a
CR2	T1	3.0	b	5.1	a	6.2	a	5.2	a	6.4	a	7.2	a	7.1	a	6.4	a
	T2	2.6	b	4.8	a	6.4	a	4.9	a	6.1	a	7.4	a	7.3	a	7.2	a
	T3	5.0	a	6.1	a	6.6	a	5.9	a	6.4	a	6.6	a	6.7	a	6.6	a
	T4	3.9	ab	5.0	a	6.3	a	5.4	a	6.1	a	7.2	a	7.2	a	6.8	a
CR3	T1	15.9	a	20.6	a	22.6	a	21.9	a	7.4	a	7.7	a	7.6	a	7.5	a
	T2	16.9	a	22.4	a	22.9	a	22.5	a	8.1	a	7.2	a	8.3	a	7.7	a
	T3	17.7	a	18.8	a	22.2	a	18.4	a	7.4	a	7.8	a	7.3	a	7.4	a
	T4	16.7	a	20.8	a	21.1	a	19.9	a	7.2	a	7.8	a	7.6	a	7.2	a

Means within one HS and CR followed by the same letter are not significantly different according to Tukey's HSD test ($\alpha=0.05$)

4.4.2 Analysis of costs

4.4.2.1 Variable costs of herbicides and machinery

The variable costs of herbicides and machinery are presented in Table 12. The variable costs for machinery in CR3, where corn was cultivated, were approximately 100 € ha-1 higher compared to the other CRs (Figure 1). However, the costs in CR2, where spring barley was cultivated, were equal to CR1. Ploughing (T2) and the deep false seedbed with rotary harrow (T4) were the most expensive tillage measures. The shallow false seedbed with harrow (T3) was the cheapest tillage measure with 19.45 € more per ha and year, compared to the conventional reduced tillage (T1).

The variable costs for herbicides were highest in HS3, where herbicide sequences or mixtures were applied, followed by HS2, where the MOA were changed annually. As more spring crops were cultivated the lower the herbicide costs were. HS4, where exclusively HRAC-A&B herbicides were applied, resulted in the lowest costs (Table 12).

Table 12. Variable costs of tillage operations (T) and herbicide strategies (HS) in the three crop rotations (CR), summarized for both experimental years.

Variable costs in € ha^{-1}	CR1	CR2	CR3
		Variable costs for machinery	
T1	412.1	412.1	517.1
T2	451.0	451.0	556.0
T3	439.0	439.0	544.0
T4	479.5	479.5	550.8
		Variable costs for herbicides	
HS1	107.9	107.9	126.0
HS2	261.0	220.6	149.2
HS3	352.0	275.8	149.2
HS4	138.8	169.2	149.2

4.4.2.2 Contribution margins (CM)

CMs as affected by CR, HS, T and year are shown in Table 13 for both locations and years. The extreme weather conditions in 2018 have affected crops yields and sales revenues negatively. The lowest CMs were achieved in the untreated control (HS1) in all CRs and at both locations. All other CMs were inconsistent and it was not possible to derive clear trends. At Ihinger Hof the low yields and the higher variable costs of HS2 and HS3 resulted in low CMs in those treatments at all CRs. Due to the higher yields at Wurmberg the highest CMs were almost always achieved in the HS3 treatment followed by HS2, independent of the CR. The cultivation of spring crops increased the CMs significantly at Wurmberg, however, this effect was not detected at Ihinger Hof. Whereas the false seedbed treatments (T3 & 4) achieved higher CMs at Ihinger Hof than the other tillage measures, ploughing (T2) and the shallow false seedbed (T3) performed better at Wurmberg. Nevertheless, the tillage measures (T2-T4) were more expensive than the conventional reduced tillage (T1), but achieved almost always higher CMs and lower blackgrass infestations in all CRs at both locations (Table 13).

Table 13. Contribution margin in € ha^{-1} in the three crop rotations (CR), four herbicide strategies (HS) and four tillage measures (T) of both locations and years.

Contribution margins in € ha^{-1}		Ihinger Hof (2017 + 2018)			
		HS1	HS2	HS3	HS4
CR1	T1	18.34	538.78	344.19	678.26
	T2	134.89	603.52	391.66	639.38
	T3	198.7	719.14	559.09	720.46
	T4	313.67	229.66	225.04	662.73
CR2	T1	-142.81	419.4	465.16	416.32
	T2	-80.73	403.11	448.87	389.39
	T3	21.63	609.09	551.23	663.14
	T4	427.53	341.42	323.42	464.55
CR3	T1	299.86	560.35	473.05	496.77
	T2	186.16	610.6	458.87	592.63
	T3	514.95	624.56	472.83	609.4
	T4	59.44	112.07	167.94	72.21
Contribution margins in € ha^{-1}		Wurmberg (2018 + 2019)			
		HS1	HS2	HS3	HS4
CR1	T1	-102.55	400.62	827.75	419.21
	T2	-193.24	482.63	1082.46	363.06
	T3	-250.32	494.63	939.03	271.44
	T4	-135.35	730.49	846.76	334.6
CR2	T1	509.6	919.04	1033.88	828.26
	T2	341.85	868.21	1069.4	897.01
	T3	828.12	945.28	996.36	962.13
	T4	537.9	834.43	1003.74	875.18
CR3	T1	672.84	887.59	943.46	897
	T2	811.37	817.28	1055.46	920.72
	T3	714.18	812.42	841.63	717.54
	T4	629.63	881.43	852.94	727.74

4.5 Discussion

The trend of blackgrass infestation development of the five previous experimental years continued. The more spring crops were included into the crop rotation, the lower the blackgrass infestations were. The blackgrass reduction was intensified, when herbicide sequences and mixtures or an annual herbicide MOA rotation were used in addition. If herbicides with the same MOA every year were used and winter annual crops were dominating the crop rotation, blackgrass abundances further increased. These results stay in in line with investigations of Gerhards et al. (2013) who determined an increase of blackgrass abundance when winter annual crops were predominant in crop rotations. Moreover, Chauvel et al. (2001) and Lutman et al (2013) demonstrated that blackgrass infestations are decreased due to the cultivation of spring barley. We assumed that spring cropping reduces blackgrass abundance, because it is a winter annual weed species which predominantly germinates in autumn (Naylor, 1972a). Further we assumed that the reduction will be enhanced as more spring crops were cultivated, because less seeds enter the soil and the soil seedbank has a short durability (Cousens and Moss, 1990). These assumptions can be stated as confirmed.

Reduced tillage resulted in the highest blackgrass densities in all years at both locations. Inversion tillage with the moldboard plough reduced infestations significantly, but densities increased again when ploughing was omitted. Other studies also observed increasing blackgrass infestations due to reduced tillage practices and decreasing infestations by ploughing (Pollard et al., 1982; Moss, 1987a; Hurle,1993; Turley et al., 1996). This might be a result of the accumulation of blackgrass seeds in the top soil due to reduced tillage, which promotes the blackgrass germination (Cussans et al., 1996). However, inversion tillage increases costs, is time consuming, can decrease the soil humus content and cause soil degradation (Power and Follet, 1987; Sepetiene and Sepetys, 2005; Moss, 2017). Moreover, Moss (2017) recommended to plough only once every five years, because the soil seed bank of blackgrass is reduced by 70-80 % per year (Cousens and Moss, 1990). Therefore, inversion tillage is not suitable for every type of arable land and it is not recommended to plough annually, even from a weed management perspective. The above-mentioned findings illustrate the demand for alternative and reliable tillage measures in years or cropping systems where ploughing is omitted.

The false seedbed preparation might be a suitable tool for blackgrass reduction. Both measures, the shallow false seedbed with the harrow and the deep with the rotary harrow, reduced blackgrass infestations up to 70 % without any additional herbicide application. In combination with spring crops and herbicide applications, the blackgrass reductions increased to 95 % and 100 %, respectively. The seed entry into the soil as well as its removal are key aspects for blackgrass management (Moss 2017). Both the shallow and the deep false seedbed preparation resulted in a fine soil texture, which is a requirement for high germination rates of blackgrass (Thurston, 1972). Further, studies have shown that besides the light stimulus, soil disturbance can affect soil temperature, water potential and changes in oxygen as well as nitrogen content which can lead to a break in dormancy of weed seeds and induce germination (Jensen, 1995; Gerhards et al., 1998). We assume that the deeply prepared seedbed was more effective than the shallow, because of the additional later sowing date, which is known to reduce blackgrass densities (Hurle, 1993; Melander, 1995; Lutman et al., 2013). Our assumption that false seedbed preparation decreases blackgrass abundance compared to sole conservation tillage practices was confirmed. Even more it performed better than ploughing, especially in year seven where ploughing was omitted.

Nevertheless, as shown in 2019 at Wurmberg, tillage effects can be non-existent if the germination process prior to seeding the crop is not induced due to missing moisture (Colbach et al., 2002b). Further, two passes with the rotary harrow turned out to be the most expensive treatment and might therefore not be appropriate for large farms and will probably not find any acceptance by the farmers. Therefore, the authors advise to use the harrow instead of the rotary harrow to induce germination of blackgrass seeds. The best application period is between mid of September to mid of October and it should be combined with a late sowing of the crop at the end of October or a spring crop cultivation. Additionally, the harrow resulted in the lowest increase of machinery costs and harrows have typically large working widths (up to 24 m), which allows a high area output in autumn when high workload peaks may occur.

Further we assumed, that the CMs of cropping systems combining different IWM strategies are higher than purely chemical weeding system. This assumption could not be confirmed. The review of Oerke (2006) has demonstrated, that abiotic factors have a higher impact on crop yields than biotic factors. In our study, the environmental conditions affected the crops yields and therefore the sale revenues the most. Even though the blackgrass abundance was significantly reduced due to spring cropping, diverse HSs and different tillage measures as well as their interactions, those IWM strategies did not necessarily result in higher crop yields. However, the higher variable costs for machinery and herbicides even did not necessarily result in lower CMs. Further, spring cropping achieved high CMs even without the use of herbicides and variable costs for herbicides decreased with increasing spring crop percentage in the CR. Additionally, when no herbicides or herbicides with the same MOA were applied and winter annual crops were predominant yields start to decrease. Therefore, it can be assumed that the implementation of IWM strategies into cropping systems will increase CMs in long-term, since environmental conditions will be balanced in long-time average. Nevertheless, further investigations for the relation between CMs and IWM strategies might be needed. Additionally, high blackgrass densities are correlated with a higher risk of resistance development, which may cause yield losses in the future (Zeller et al., 2019, submitted). Gerhards et al. (2016) calculated higher CMs for IWM strategies if herbicide resistant blackgrass has already been developed. However, they stated that farmers are often reluctant to implement IWM strategies into their cropping systems and predominantly rely on herbicides, because the farmers underestimate the costs of herbicide resistance in the long-term.

4.6 Conclusion

Balanced IWM strategies should include several non-chemical weed control measures to control problematic weed species. Diverse crop rotations, tillage measures and their combinations are suitable options to decrease weed infestations. Further, due to the cultivation of diverse crops more herbicide AIs become available and if weed infestations are lower also weaker AIs can be utilized. This simplifies a MOA rotation and allows an easier chemical control of problematic weeds species, which are prone to resistance development. The authors presume, that problematic weed species such as blackgrass can solely be controlled in long-term by combining preventive and direct weed control measures. If one control option fails, another one can compensate for it. Other possible preventive measures might be inter- and cover-cropping, higher seed rates of crops, reduced row spacing and competitive crop cultivars.

4.7 Acknowledgements

The authors thank Mr. Schumacher from the Department of Weed-Science at University of Hohenheim for reviewing the manuscript. We are also grateful to Mr. Capezzone from the Biostatistics Department and Mr. Christian Sponagel from the Department of Farm Management at University of Hohenheim for their assistance regarding data analysis. Further, we also like to thank Hartmut Weeber, the colleagues at Ihinger Hof Research Station and all the students and technical assistants for their support to realize the experiments and gather data. Study was funded by the German Federal Office for Agriculture and Food.

Chapter V

–

General Discussion

5 General Discussion

In the present thesis, IWM strategies were investigated to determine control methods for herbicide resistant *A. myosuroides*. Field experiments and greenhouse bioassays were conducted over a seven-year period. The results were presented in detail in three peer-reviewed papers, which are illustrated in the chapters II – IV of the thesis. In this chapter, the most important results are summarized, discussed and specific recommendations are given for practical implementation.

Resistance to herbicides is an increasing problem in many weed species worldwide and can be attributed to an extensive use of herbicides and a resulting selection pressure. The "Herbicide Resistance Action Committee" and the "Weed Science Society of America" were founded or rather formed task forces, with the aim to find reliable solutions and management strategies for the control of herbicide resistant weeds in a global point of view (Heap, 1997; Powles and Yu, 2010; Massa, 2012). *Poaceae* are most affected with confirmed resistance cases in 82 species worldwide (Heap, 2019). Herbicide resistant *A. myosuroides* is one of the most problematic *Poaceae* of autumn sown crops in Europe. Resistant populations are reported in 14 countries. Multiple herbicide resistant biotypes in which up to five MOAs are affected, were confirmed in Belgium, Denmark, France, Germany, Netherlands, Poland, Spain, Sweden and the United Kingdom (Heap, 2019). The predominant reliance on herbicides, combined with high proportions of autumn sown crops, early sowing dates of winter cereals, widely switched cropping systems to reduced tillage and a limited use of mechanical weed control measures, led to resistances. Most affected are post-emergence herbicides, especially with regard to acetyl-CoA carboxylase (ACCase) and acetolactate synthase (ALS) inhibiting herbicides

(Cussans et al., 1979; Moss, 1987a; Hurle, 1993; Moss et al., 2007; Gerhards et al., 2013; Moss, 2017). The ACCase enzyme catalyzes the first step of the fatty acid synthesis and the ALS enzyme the first step of the biosynthesis of branched-chain amino acids valine, leucine and isoleucine (Stetter, 1994; Délye et al., 2005). Investigations of the target-site in resistant plants have shown that the gene position Ile-1781 is most commonly affected in ACCase and, Pro-197 as well as Trp-574 in ALS (Marshall and Moss, 2008; Délye et al., 2010; Marshall et al., 2013; Knight, 2015). However, also other resistance mechanisms can be responsible for herbicide resistance and especially NTSR mechanisms gain more and more in importance and are far less investigated (Drobny et al., 2006; Heap, 2014). The declining efficacy of post-emergence herbicides due to herbicide resistance, has led to an increased use of pre-emergence herbicides, which are known to be less prone to resistance development. Nevertheless, resistant populations have also been confirmed for those MOAs, mainly associated with NTSR mechanisms. In 2019, *A. myosuroides* populations are known with confirmed resistance towards seven (HRAC – A, B, C1, C2, K1, K3 and N) herbicide MOAs (Moss, 2017; Heap, 2019). In addition to the progressing resistance development, the efficacy of pre-emergence herbicides is highly dependent on environmental conditions such as adequate soil moisture. Therefore, relying on pre-emergence herbicides to control *A. myosuroides* is doubtful and cannot be considered as a reliable method in a longer-term (Moss 2017).

Compared to herbicides, many non-chemical control strategies are more cumbersome and weed control efficacy is lower. Additionally, these measures are often time consuming, labor intensive, more expensive, less predictable and cannot replace herbicides on most farms. Further, they can cause negative environmental effects such as soil degradation and the effect is often difficult to detect, especially when the used herbicides still achieve high efficacies (Moss et al., 2010). However, as it is European

Union policy to reduce pesticides in a measured way and because of the declining number of available AIs, it is not only advisable to implement IWM strategies it is seen as necessity (EU, 2009; Lefebvre et al., 2015; Drobny, 2016; Moss, 2017).

5.1 Chemical control

The investigation of different herbicide strategies has shown, that in susceptible *A. myosuroides* populations resistances developed rapidly, when herbicides with the same MOAs were continuously applied. *A. myosuroides* control efficacy (ACE) started to decrease in the field already four years after experimental start. After 5 years the susceptible populations had shifted to a resistant and significant yield losses were observed. In the greenhouse bioassay efficacy losses of up to 60 % were determined for some herbicides. These results are in line with investigations of Neve and Powles (2005) as well as Powles and Yu (2010), who also confirmed a rapid selection of resistant biotypes when a continuous application of herbicides with the same MOAs has taken place or when herbicides were used constantly at lower than the recommended rates. Herbicide resistance was confirmed by performing molecular genetic analysis. For ACCase-inhibitors up to 52 % of the surviving plants have shown mutations in the codon positions 1781 – Ile, 2027 – Trp and 2041 – Ile. Regarding ALS-inhibitors, mutations in the codon positions 197 – Pro and 574 – Trp were determined in up to 26 % of the surviving plants. Averaged over all locations and populations, approximately in 25 % of the plants, which survived the herbicide application, a mutation in the target site was found. The survival of the other 75 % must be justified by other resistance mechanisms. These observations are in line with studies of Petersen and Wagner (2009) and Rosenhauer et al. (2013) who demonstrated the importance of TSR on German arable lands and increasing infestation rates associated with TSR from approximately 10 % in 2004 on up to 54.3 % in 2011. Further, Drobny et al (2006) illustrated a gaining importance of NTSR mechanisms in *A. myosuroides* populations.

Therefore, the combination of different MOA within the cultivation of a crop and during a crop rotation is particularly recommended in an IWM perspective (Beckie, 2006; Beckie and Reboud, 2009). Plant protection consultants generally recommend herbicide-sequences or -mixtures. Sequences often combine different MOAs in pre- and post-emergence herbicide applications. They are frequently applied in winter cereals, as pre-emergence use at crop BBCH 0 – 10 in autumn and as post-emergence use in spring at BBCH 20 – 35. Mixtures cover, tank mixtures of various herbicides with different MOAs, co-packs or formulated herbicides consisting of AIs with different MOAs. Mixtures are often applied in spring crops such as sugar beet (*Beta vulgaris* L.) or corn (*Zea mays* L.). In our studies, mixtures and sequences performed well in the field and were able to achieve high ACEs over the entire experimental time. Nevertheless, a greenhouse bioassay and a following TSR-analysis demonstrated that resistances evolved, resulting in efficacy losses of up to 50 % for some AIs. Also, Hicks et al. (2018) could demonstrate a resistance development when using herbicide-mixtures and -sequences in a long-term. This observation highlights the need for reliable herbicide management strategies, which maintain the efficacy of AIs as long as possible.

Therefore, an approach was investigated in this study to use every MOA or at least chemical – family only once over a period of five years. It was aimed to reduce the selection pressure as much as possible. The possibility of higher infestations was ventured, by using weaker AIs or partially only pre-emergence herbicides, to allow a consequent rotation of herbicide MOA. The herbicide MOA rotation, has shown similar ACEs compared to the herbicide-mixtures and -sequences in field. In some years the

MOA rotation performed significantly better. When *A. myosuroides* infestations increased due to the use of weaker AIs or the sole use of pre-emergence herbicides, no significant yield loss was determined and in the following year the ACE increased again. Nevertheless, the assessment of herbicide resistance demonstrated, that there was still a selection of resistant biotypes. The highest efficacy loss of 27 % was measured for plants originating from crop rotations where exclusively winter annual crops were cultivated. This indicates that an *A. myosuroides* management relying exclusively on herbicides is no sufficient solution and underlines the demand for non-chemical control methods.

5.2 Preventive measures

In the present thesis, the effects of diverse crop rotations with different proportions of spring crops combined with cover cropping were investigated. Winter wheat (*Triticum aestivum* L.), winter oilseed rape (*Brassica napus* L.), corn and spring barley (*Hordeum vulgare* L.) were cultivated. By the continuous cultivation of winter annual crops, *A. myosuroides* densities increased from 14 (±1.6 SD) to 5347 (±264 SD) and 201(±125 SD) to 3424 (±253 SD) heads m^{-2} within 5 years at the experimental sites, when no herbicides were used and reduced tillage was performed. Due to the cultivation of cover crops followed by spring crops infestations were significantly reduced. Densities were on average 72 % lower in years when spring crops were cultivated. These results are in line with investigations of Naylor (1972a) and Blair et al. (1999), who demonstrated the high preference of *A. myosuroides* to winter annual crops, especially cereals. Further, Chauvel et al. (2001) and Lutman et al (2013) described the positive effect on *A. myosuroides* infestations by spring cropping. Also cover-cropping is known to suppress *A. myosuroides* infestations (Brust et al., 2014). Whereas corn and spring barley reduced *A. myosuroides* densities significantly, spring barley was more resistive. The lower *A. myosuroides* densities resulted in a lower seed entry into the soil and lower infestations in the subsequent years. Compared to a crop rotation where 100 % winter annual crops were cultivated, a proportion of 25 % or 50 % spring crops in the crop rotation resulted in a lower infestation of averaged 20 % or 41 % after 5 years, respectively. Colbach et al. (2006, 2010) estimated similar effects by using the ALOMYSYS prediction model, which was developed to forecast *A. myosuroides* emergence.

Lower *A. myosuroides* infestations are accompanied with a lower seed content in the soil and generate a basis for the chemical management with herbicides. A higher infestation results in a higher risk for the occurrence of genetically modified biotypes. Due to the use of herbicides, these biotypes will be selected, especially when herbicides with the same MOAs are frequently applied (Powles and Yu, 2010; Hicks et al., 2018). The investigations of this thesis confirmed the assumptions. The cultivation of 100 % winter annual crops and a high selection pressure by herbicides resulted in 917 *A. myosuroides* heads m^{-2} after five years. By the cultivation of 75 % winter annual crops, *A. myosuroides* heads m^{-2} amounted to 770 and 366 heads m^{-2} were counted when the crop rotation had a proportion of 50 % winter annual crops. At the other location, *A. myosuroides* heads m^{-2} amounted to 1850, 1138 and 477, respectively. Additionally, an investigation of TSR has shown, that multiple mutations within one sample were found, especially for plants originating from crop rotations with 100 % winter annual crops and when high selection pressure was exerted by herbicides. Otherwise, the results have shown, that solely by the combination of an annual change of herbicide MOAs and crop rotations with 50 % spring crops, the development of resistance was inhibited. It was even possible to reduce the frequency within an already established resistant population at one of the sites, through the combination of these IWM measures.

These findings highlight the need to combine chemical and preventive measures to suppress *A. myosuroides* effectively and maintain Ais as an important component in a resistance management perspective. Wider and diverse crop rotations allow the use of more MOAs, chemical families and AIs. Herbicides such as propyzamid (HRAC – K1), aclonifen (F3), metribuzin (C1) and cyclohexadiene (A) are not available in cereals and can exclusively be used in diverse crop rotations. Additionally, lower infestations enable the use of weaker AIs and due to the cultivation of crops such as corn, sugar beet and potatoes (*Solanum tuberosum* L.) mechanical weeding for instance hoeing can be carried out.

5.3 Mechanical measures

On the basis of its biology, *A. myosuroides* is strongly promoted by reduced tillage. Reduced tillage measures accumulate seeds in the top soil. As the light stimulus is particularly important for germination, 90 % of the seedlings germinate from soil layers not deeper than 0.05 m (Naylor, 1972a; Moss, 1985; Cussans et al., 1996; Colbach et al., 2002a; Anderson and Espeby, 2008). Additionally, *A. myosuroides* seeds have a high germination rate and 60 – 70 % of an infestation originate from fresh seeds not older than one year (Naylor, 1972b; Froud-Williams, 1985). Therefore, densities can increase rapidly in reduced tillage systems, whilst infestations can significantly be reduced by moldboard ploughing (Pollard et al., 1982; Moss, 1987a; Hurle, 1993; Turley et al., 1996; Lutman et al., 2013). In these experiments, *A. myosuroides* infestations were on average reduced by 49 % due to moldboard ploughing. By the combination with spring crops control efficacies were increased. Nevertheless, moldboard ploughing is an expensive and time-consuming tillage measure, which farmers like to omit. Additionally, the frequent use results in a decreasing soil humus content and can lead to soil degradation (Lindstrom et al., 1992; Slepetiene and Slepetys, 2005; KTBL, 2019). Further, Cousens and Moss (1990) demonstrated that buried seeds lose annually 70 – 80 % of their fertility. Therefore, it is not advisable to plough annually, neither from soil fertility nor from weed management perspective. However, it was shown that in years where ploughing was omitted *A. myosuroides* infestations increased rapidly.

Because of the germination characteristics and because *A. myosuroides* is solely propagated by seeds, the seed entry into the soil and its removal are possibilities to suppress infestations with intensive tillage measures in autumn. Therefore, a shallow false seedbed with a harrow and a deep false seedbed with a rotary harrow prior to seeding were tested. Results have demonstrated an approximate *A. myosuroides* decrease of 30 % per year, when two treatments with the harrow were carried out in mid and end of September. The deep measure with the rotary harrow reduced blackgrass abundance up to 70%. The reduction of the blackgrass infestations was increased to 95 % and 100 %, if those measures were combined with spring crops and alternating herbicide MOA, respectively. Both measures resulted in a very fine soil texture, which is important for *A. myosuroides* germination (Thurston, 1972). Further, the soil disturbance affects the water potential, soil temperature and can change the oxygen as well as nitrogen content, which can break seed dormancy of weed seeds and induce germination (Jensen, 1995; Gerhards et al., 1998). It can be assumed, that the deep false seedbed was more effective, because of the additional later seeding date. Later sowing dates of winter cereals are a further preventive measure known to reduce blackgrass abundance effectively (Hurle, 1993; Melander, 1995; Lutman et al., 2013). This highlights the benefits of combining

IWM strategies. Nevertheless, both measures performed better than ploughing and might be a solution for cropping systems where ploughing is omitted.

However, it was also shown, that the efficacies of pre-cultivation tillage measures can be decreased or rather fail to appear completely. For example, if there is no precipitation between the tillage treatments and not enough moisture in the soil, it is likely that the germination of weeds and consequently the effects stay off (Colbach et al. 2002b). Further, yields can be reduced by delayed sowing, when wet and cold conditions occur directly after sowing or by pre-emergence herbicide applications in the late season (Metcalf et al., 1970; Freyman et al., 1982; Ferreira et al., 1990). Moreover, intensive tillage measures in autumn stay in contrast to a cover crop cultivation. Cover crops are typically cultivated before spring crops and sowing is performed as soon as possible after harvest (Sturm et al., 2017). Additionally, agricultural direct payments of the European Union policy are related to "Greening" measures. Thus, the cultivation of cover crops is an essential part for farmers to receive direct payments.

5.4 Analysis of costs

The analysis of costs has shown, that the variable costs increased due to the cultivation of corn. However, the costs for spring barley and winter wheat cultivation were equal and both spring crops reduced *A. myosuroides* abundance significantly. The more spring crops were cultivated the lower the variable costs for the herbicide application were. The conventional reduced tillage measure was the cheapest but always resulted in highest *A. myosuroides* infestations. Ploughing and the deep false seedbed increased costs the most and were time consuming. Both, Power and Follet (1987) and Heap (2014), thus declared the dominance of reduced tillage systems, which predominantly omit mechanical treatments and rely on herbicides. The lowest increase of costs was achieved due to the shallow false seedbed with the harrow. The costs per ha and year were 19.45 € higher.

Even though the higher variable costs it was assumed, that the implementation of IWM strategies will increase the contribution margins (CMs), because all of those measures suppressed *A. myosuroides* significantly. However, the environmental conditions affected the crop yields and sale revenues the most and it was not possible to derive clear trends. These results stay in line with the review of Oerke (2006) who has demonstrated, that abiotic factors have the highest impact on crop yields. Nevertheless, lowest CMs were always achieved when no herbicides were applied. Even though all tested tillage measures were more expensive than conventional reduced tillage, they almost always achieved higher CMs. It was demonstrated, that the predominant cultivation of winter annual crops enhanced *A. myosuroides* abundance and herbicide resistance development. Due to the use of herbicides with the same MOA, resistant biotypes were rapidly selected. Gerhards et al. (2016) calculated lower CMs when herbicide resistance increases in a long term. Therefore, it can be assumed, that the combination of different IWM strategies will increase CMs in long-term. Especially, if environmental conditions will be balanced due to the long-time average. Nevertheless, further investigations between the relation of IWM strategies and CMs might be needed.

5.5 Recommendation

An annual ACE of 95 % is necessary to prevent an increase of *A. myosuroides* populations (Moss, 2017). This thesis demonstrated, that even by low selection pressure resistant biotypes might be selected and resistance may evolve. Therefore, management systems relying solely on herbicides are doubtful in a longer-term, even if a MOA rotation is well performed. Nevertheless, it was shown that the combination with preventive measures was able to slow down or even prevent resistance development and reduce already established resistances. Due to the combination with other strategies like rotational ploughing, false seedbed preparation and delayed sowing *A. myosuroides* infestations and resistance development can be further increased. CMs are not necessarily lower because of the higher costs of combining different IWM strategies, especially considering the positive impact on resistance development. Since *A. myosuroides* is relatively immobile, the avoidance and the management of resistant biotypes is to a great extent within the farmer´s own disposition to apply and combine IWM strategies. A rethinking of farmers would be important as fast solutions are unlikely to be expedient and a reliance solely on herbicides is no alternative (Moss et al., 2007; Moss, 2017). Only the combination of preventive, curative, mechanical and chemical weed control tactics, can achieve a firm long-term management and enable a balance even if some measures fail. Nevertheless, many farmers are reluctant to change their cropping system and to implement IWM measures, because they underestimate the arising costs due to herbicide resistance and yield loss in long-term (Gerhards et al. 2016).

VI – Summary

–

VII – Zusammenfassung

6 Summary

Due to the growing world population, there is an increasing demand for agricultural products. Agricultural yields are mainly limited by abiotic and biotic factors. Weeds are the most important biotic factor and cause highest yield losses, worldwide. Since the discovery and introduction of herbicides, weeds were mainly controlled chemically. Herbicides achieve high control efficacies, are selective and relatively cost-efficient. Therefore, many other weed control measures have been neglected and replaced by the exclusive use of herbicides. However, the intensive use of herbicides led to a selection of resistant weed species worldwide.

Alopecurus myosuroides Huds. (blackgrass) is one of the most important weeds in Europe and herbicide resistance is an increasing problem. At the same time, more and more active ingredients are losing their permission and are no longer available for *A. myosuroides* control. High *A. myosuroides* infestations result in increasing yield losses. Integrated weed management strategies (IWM) combine different measures to enhance control efficacies, prevent the development of herbicide resistances and secure crop yields in long-term.

The aim of this dissertation was to investigate the influence of diverse IWM strategies on *A. myosuroides*. For this purpose, field and greenhouse experiments were conducted to test:

1. different chemical strategies with varying selection pressure,
2. various preventive and curative measures, such as crop rotation, cover cropping and delayed sowing time,

3. diverse mechanical measures, which mainly consisted of soil tillage performed between crop cultivation,
4. the interactions of these chemical, preventive, curative and mechanical measures.

Primarily, the parameters population– and resistance development of *A. myosuroides* as well as the influence on crop yield were investigated. Further the contribution margins of the different cropping systems were evaluated.

The results have shown, that susceptible populations can develop resistances rapidly. Herbicide strategies, which are currently recommended and used in practice, resulted in high resistances. Even due to minimal selection pressure, resistant biotypes were selected. Preventive and curative measures, were able to reduce *A. myosuroides* infestations and resistance development significantly. Exclusively the exertion of low selection pressure combined with the performance of preventive and curative measures was able to prevent resistance development. There was a synergistic effect between the tillage measures and all other tested measures resulting in an increased control efficacy. Due to combining different IWM strategies the costs of the cropping system increased, but the increasing costs had not necessarily a negative effect on the contribution margins. The results demonstrated, that a long-term *A. myosuroides* control will be exclusively possible by combining preventive, curative, mechanical and chemical measures. Therefore, the current agricultural practice should be reconsidered.

7 Zusammenfassung

Mit einer wachsenden Weltbevölkerung geht ein steigender Bedarf an landwirtschaftlichen Produkten einher. Landwirtschaftliche Erträge werden hauptsächlich durch abiotische und biotische Faktoren limitiert. Weltweit sind Unkräuter der am stärksten limitierende biotische Faktor. Seit der Entdeckung von Herbiziden, werden Unkräuter vorwiegend chemisch bekämpft. Herbizide erzielen hohe Wirkungsgrade, sind spezifisch einsetzbar und relativ kostengünstig. Daher wurden viele andere unkrautregulierende Maßnahmen vernachlässigt und durch den ausschließlichen Einsatz von Herbiziden ersetzt. Der intensive Herbizid-gebrauch führte jedoch weltweit zu einer Selektion resistenter Unkrautarten.

Alopecurus myosuroides Huds. (Ackerfuchsschwanz) ist eines der bedeutendsten Ackerungräser in Europa und zunehmend von Resistenzbildungen betroffen. Gleichzeitig verlieren mehr und mehr Wirkstoffe ihre Zulassungen und stehen somit nicht mehr zur Bekämpfung von *A. myosuroides* zur Verfügung. Der steigende Befall von Ackerflächen durch *A. myosuroides* führt vielerorts zu steigenden Ertragsverlusten. Strategien zur integrierten Unkrautbekämpfung kombinieren nicht-chemische und chemische Maßnahmen, um Wirkungsgrade zu steigern, Resistenzbildungen vorzubeugen und Erträge langfristig zu sichern.

Zielsetzung dieser Dissertation war es, den Einfluss verschiedener Strategien zur integrierten Unkrautbekämpfung auf *A. myosuroides* zu untersuchen. Hierfür wurden Feld- und Gewächshausversuche durchgeführt, um:

1. verschiedene chemische Strategien mit variierendem Selektionsdruck,

2. verschiedene präventive und ackerbauliche Maßnahmen, wie Fruchtfolge, Zwischenfruchtanbau und Aussaatzeitpunkt,
3. verschiedene mechanische Maßnahmen, welche vor allem die Bodenbearbeitung zwischen den Kulturen umfassten,
4. die Interaktionen dieser chemischen, präventiven, ackerbaulichen und mechanischen Maßnahmen,

zu untersuchen. Dabei wurden vornehmlich die Parameter Populations- und Resistenzentwicklung von *A. myosuroides* sowie deren Einfluss auf die Ertragsparameter der Kulturpflanzen beurteilt. Außerdem wurden die Deckungsbeiträge der unterschiedlichen Anbauverfahren ermittelt.

Die Ergebnisse haben gezeigt, dass sensitive Populationen binnen kürzester Zeit Resistenzen bilden können. Auch Herbizidstrategien, welche derzeit in der Praxis empfohlen und angewandt werden, resultierten in starken Resistenzbildungen. Selbst durch minimalen Selektionsdruck wurden resistente Biotypen selektiert. Durch präventive und ackerbauliche Maßnahmen, konnte der Befall durch *A. myosuroides* sowie eine Resistenzentwicklung erheblich reduziert werden. Ausschließlich durch eine Kombination von niedrigem Selektionsdruck sowie präventiven und ackerbaulichen Maßnahmen konnte eine Resistenz-entwicklung gänzlich verhindert werden. Die unterschiedlichen Boden-bearbeitungen wirkten synergistisch und steigerten die Effektivität der anderen getesteten Maßnahmen erheblich. Durch eine Kombination von mehreren Maßnahmen zur integrierten Unkrautbekämpfung wurden die Kosten zwar erhöhten, jedoch waren die Deckungsbeiträge nicht zwangsläufig niedriger. Die Ergebnisse haben gezeigt, dass eine langfristige Regulierung von *A. myosuroides* ausschließlich durch eine Kombination von präventiven, ackerbaulichen, mechanischen und chemischen Maßnahmen möglich sein wird und in der praktischen Landwirtschaft ein Umdenken erfolgen sollte.

Chapter VIII

–

References

8 References

Agraronline, 2019. Pesticides 2019. *Available at https://www.myagrar.de (last accessed 15.10.2019).*

Agrolandis, 2015. Prices for winter cereal seed. *Available at http://www.agrarservice-repenning-specht.de (last accessed 15.10.2019).*

Alexandratos, N., Bruinsma, J., 2012. World agriculture towards 2030/2050: the 2012 revision. *In: ESA Working paper No. 12-03. Rome, FAO.*

Anderson, L., Espeby, L.Å., 2008. Variation in seed dormancy and light sensitivity in *Alopecurus myosuroides* and *Apera spica-venti. Weed Res. 49, 261–270.*

BayWa, 2019. RAGT seeds spring barley Planet. *Available at https://www.baywa.de (last accessed 15.10.2019).*

Beckie, H.J., 2006. Herbicide-Resistant Weeds: Management Tactics and Practices. *Weed Tech. 20, 793-814.*

Beckie, H.J., Reboud, X., 2009. Selecting for Weed Resistance: Herbicide Rotation and Mixture. *Weed Tech. 23, 363-370.*

Blair, A.M., Cussans, J.W., Lutman, P.J.W., 1999. A biological framework for developing a weed management support control in winter wheat. *In: 1999 Brighton Conference – Weeds. Brighton, UK, 753-760.*

Brust, J., Claupein, W., Gerhards, R., 2014. Growth and weed suppression ability of common and new cover crops in Germany. *Crop Prot. 63, 1-8.*

Buhler, D.D., 2002. 50th Anniversary – Invited Article Challenges and opportunities for integrated weed management. *Weed Sci, 50, 273 – 280.*

Campe, R., Hollenbach, E., Kämmerer, L., et al. 2018. A new herbicidal site of action: Cinmethylin binds to acyl-ACP thioesterase and inhibits plant fatty acid biosynthesis. *Pestic. Biochem. Physiol. 148, 116–125.*

Chauvel, B., Guillemin, J.P., Colbach, N., Gasquez, J., 2001. Evaluation of cropping systems for management of herbicide-resistant populations of blackgrass (*Alopecurus myosuroides* Huds.). *Crop Prot. 20, 127-137.*

Christen, O., Fried, W., 2007. Winter rape. *The handbook for professionals. Ed. DLG-publisher, Frankfurt, Germany. 323.*

Colbach, N., Chauvel, B., Duerr, C., Richard, G., 2002a. Effect of environmental conditions *Alopecurus myosuroides* germination. I. Effect of temperature and light. *Weed Res 42, 210-221.*

Colbach, N., Duerr, C., Chauvel, B., Richard, G., 2002b. Effect of environmental conditions *Alopecurus myosuroides* germination. II. Effect of moisture conditions and storage length. *Weed Res 42, 222-230.*

Colbach, N., Duerr, C., Estrade, R.J., Chauvel, B., Caneill, J., 2006. AlomySys: Modelling black-grass (*Alopecurus myosuroidesv* Huds.) germination and emergence, in interaction with seed characteristics, tillage and soil climate I. Construction. *J. Agron. 24, 95–112.*

Colbach, N., Chauvel, B., Gauvrit, C., Munier-Jolain N.M., 2007. Construction and evaluation of ALOMYSYS modelling the effects of cropping systems on the blackgrass life-cycle: From seedling to seed production. *Ecol. Modelling 201, 283–300.*

Colbach, N., Schneider, A., Ballot, R., Vivier, C., 2010. Diversifying cereal-based rotations to improve weed control. Evaluation with the ALOMYSYS model quantifying the effect of cropping systems on a grass weed. *OCL 17, 292 – 300.*

Cousens, R., Moss, S.R., 1990. A model of the effects of cultivation on the vertical distribution of weed seeds within the soil. *Weed Res. 30, 61-70.*

Cussans. G.W., Moss, S.R., Pollard, F., Wilson, B.J., 1979. Studies of the effect of tillage on annual weed populations. *In Proceedings European Weed Research Society Symposium, the influence of different factors on the development and control of weed, Mainz, Germany. 115-122.*

Cussans, G.W., Raundonius, S., Brain, P., Cumberworth, S., 1996. Effects of depth of seed burial and soil aggregate size on seedling emergence of Alopecurus myosuroides, Galium aparine, Stellaria media and wheat. *Weed Res. 36, 133-141.*

De Prado, R.A., Franco, A.R., 2004. Cross-resistance and herbicide metabolism in grass weeds in Europe: biochemical and physiological aspects. *Weed Sci. 52, 441-447.*

Délye, C., Zhang, X.Q., Michel, S. et al. 2005. Molecular Bases for Sensitivity to Acetyl-Coenzyme A Carboxylase Inhibitors in Black-Grass. *Plant Physiol. 137, 794–806.*

Délye. C., Michel, S., Bérard, A., 2010. Geographical variation in resistance to acetyl-coenzyme A carboxylase-inhibiting herbicides across the range of the arable weed *Alopecurus myosuroides* (black-grass). *New Phytol. 186, 1005–1017.*

Délye, C., 2013. Unravelling the genetic bases of non-target-site-based resistance (NTSR) to herbicides: a major challenge for weed science in the forthcoming decade. *Pest Manag. Sci. 69, 176-187.*

Délye, C., Jasieniuk, M., Le Corre, V., 2013. Deciphering the evolution of herbicide resistance in weeds. *Trends Genet. 29, 649–658.*

Devine MD, Shukla A. 2000. Altered target sites as a mechanism of herbicide resistance. *Crop Prot. 19, 881–889.*

Drobny, H.G.; Salas, M.; Claude, J.P., 2006. Management of metabolic resistant black-grass (*Alopecurus myosuroides* Huds.) populations in Germany—Challenges and opportunities. *J. Plant Dis. Prot., 25, 65–72.*

Drobny, H.G., 2016. Weed management under pressure. Is it possible to control every weed in the future with herbicides? *In: 27th German Conference on Weed Biology and Weed Control, February 23–25, Braunschweig, Germany. Julius-Kühn-Archiv 452, 19-23.*

Duke, S.O., 2011. Why have no new herbicide mode of action appeared in recent years? *Pest Manag. Sci. 68, 505–512.*

EU., 2009. Directive 2009/128/EC of the European Parliament and of the Council of 21 October 2009 establishing a framework for Community action to achieve the sustainable use of pesticides. *Available at: http://eur-*

lex.europa.eu/LexUriServ/LexUriServ.do?uri=OJ:L:2009:309:0071:0086:en: PDF (Accessed 2019 /09 / 17).

FAO, 2006. World agriculture: towards 2030/2050 – Interim report. Rome. *Available at: http://www.fao.org/fileadmin/user_upload/esag/docs/ Interim_report_AT2050web.pdf (Accessed 2019/09/02).*

Ferreira, K.L., Baker, T.K., Peeper, T.F., 1990. Factors Influencing Winter Wheat (*Triticum aestivum*) Injury from Sulfonylurea Herbicides. *Weed Tech. 4, 724-730.*

Foster, D., Ward, P., Hewson, R., 1993. Selective grass-weed control in wheat and barley based on the safener fenchlorazole-ethyl. *In: Brighton Crop Protection Conference - Weeds. Surrey, Great Britain. 1267-1267.*

Freckleton, R.P., Hicks, H.L., Comont, D., et al., 2018. Measuring the effectiveness of management interventions at regional scales by integrating ecological monitoring and modelling. *Pest Manag. Sci. 74, 2287–2295.*

Freyman, S., Moyer, J.R., Schaalje, G.B., 1982. Effect of herbicides on cold hardiness of orchardgrass. *Can. J. Plant Sci., 62 (3), 801-804.*

Fritzsche, R., Seemann, E., Werner, B., Mol, F., Gerowitt, B., 2012. Gaining extra information from herbicide trials - analysing field trials of the region Hannover from 2003 to 2009. *In: 25th German Conference on Weed Biology and Weed Control, March 13–15, Braunschweig, Germany. Julius-Kühn-Archiv 434, 409-416.*

Froud-Williams, R., 1988. Changes in weed flora with different tillage and agronomic management systems. *In Weed management in agroecosystems: ecological approaches. Altieri, M.A., Liebman, M., CRC. Boca Raton, Florida. USA. 213-236.*

Gaines, T.A., Zhang, W., Wang, D., et al. 2010. Gene amplification confers glyphosate resistance in *Amaranthus palmeri. Proc. Natl. Acad. Sci. USA 107 (3), 1029-1034.*

Gerhards, R., Juroszek, P., Kluemper, H., Kuehbauch, W., 1998. Possibilities for weed management on field sites due to photo-biological control. *Crop Sci. 2 (2), 91-96.*

Gerhards, R., Dieterich, M., Schumacher, M., 2013. Decrease of weed species diversity in Baden-Wuerttemberg - a comparison of vegetation surveys in 1948/49, 1975-1978 and 2011 in the area of Mehrstetten - recommendations for agriculture and nature conservation. *Healthy Plant 65, 151-160.*

Gerhards, R., Dentler, J., Gutjahr, C., et al., 2016. An approach to investigate the costs of herbicide resistant *Alopecurus myosuroides* Huds. *Weed Res. 56, 407–414.*

Gerowitt, B., Heitefuss, R., 1990. Weed economic thresholds in cereals in the Federal Republic of Germany. *Crop Prot. 9, 323–331.*

Gressel, J., Segel, L.A., 1990. Modelling the effectiveness of herbicide rotations and mixtures as strategies to delay or preclude resistance. *Weed Tech. 4, 186–198.*

Heap, I.M., 1997. The occurrence of herbicide-resistant weeds worldwide. *Pestic Sci 51 (3), 235-243.*

Heap, I., 2005. Criteria for Confirmation of Herbicide-Resistant Weeds with specific emphasis on confirming low level resistance. *Available at: http://www.hracglobal.com (accessed 09/09/2019).*

Heap, I., 2014. Global perspective of herbicide-resistant weeds. *Pest Mana. Sci. 70, 1306-1315.*

Heap, I., 2019. International survey of herbicide resistant weeds. *Available at: http://www.weedscience.org (accessed 09/09/2019).*

Hicks, H.L., Comont, D., Coutts, S.R., et al., 2018. The factors driving evolved herbicide resistance at a national scale. *Nat. Ecol. Evol. 2, 529.*

Hurle, K., 1993. Integrated management of grass weeds in arable crops. *In: Proceedings Brighton crop protection conference, weeds. Brighton, UK, 22-25 November. 81-88.*

Jensen, P.K., 1995. Effect of light environment during soil disturbance on germination and emergence pattern of weeds. *Ann. Appl. Biol. 127, 561-571.*

Jutsum, A.R., Graham, J.C., 1995. Managing weed resistance: the role of the agrochemical industry. *In: Proceedings of the Brighton Crop Protection Conference-Weeds. Farnham, U.K.: 783-790.*

Knab, W., Hurle, K., 1988. Effect of tillage on blackgrass (*Alopecurus myosuroides* Huds.). *J. Plant Dis. Prot. 11, 97–108.*

Knight, C.M., 2015. Investigating the evolution of herbicide resistance in UK populations of *Alopecurus myosuroides. PhD thesis, University of Warwick.*

Horowitz, M., 1975. Application of bioassay techniques to herbicide investigations. Weed Res. 16. 209-215.

KTBL 2019. KTBL-Fieldwork calculator. *Available at: https://daten.ktbl.de/feldarbeit/home.html (last accessed 09/05/2019)*

Lagator, M., Vogwill, T., Mead, A., et al., 2013. Herbicide mixtures at high doses slow the evolution of resistance in experimentally evolving populations of Chlamydomonas reinhardtii. *New Phyt. 198, 938-945.*

Lefebvre, M., Langrell, S.R., Gomez-y-Paloma, S., 2015. Incentives and policies for integrated pest management in Europe: a review. *Agron. Sust. Develop. 35 (1), 27-45.*

Leighty, C.E., 1938. Crop rotation. *In: Soils and men: USDA Yearbook of Agriculture. U.S. Government Printing Office, Washington, D.C., USA, 406-430.*

LEL, 2019. State Institute for Agriculture, Food and Rural Areas Schwäbisch Gmünd. *Available at www.lel-bw.de (last accessed 15.10.2019).*

Lindstrom, M.J., Nelson, W.W., Schumacher, T.E., 1992. Quantifying tillage erosion rates due to moldboard plowing. *Soil and Tillage Res. 24 (3), 243-255.*

LTZ, 2019. State Institute for Agrar Technology Augustenberg. *Available at www.ltz.landwirtschaft-bw.de (last accessed 22.11.2019).*

Lutman, P.J.W., Moss S.R., Cook, S., Welham, S.J., 2013. A review of the effects of crop agronomy on the management of *Alopecurus myosuroides. Weed Res. 53, 299-313.*

Marshall, R., Moss, S.R., 2008. Characterization and molecular basis of ALS inhibitor resistance in the grass weed *Alopecurus myosuroides. Weed Res. 48, 439–447.*

Marshall, R., Hanley, S., Hull, R., Moss, S.R., 2013. The presence of two different target site resistance mechanisms in individual plants of *Alopecurus myosuroides*

Huds. identified using a quick molecular test for the characterisation of six ALS and seven ACCase SNPs. *Pest Manag. Sci. 69, 727–737.*

Massa, D., 2012. Investigations on herbicide resistance in *Apera spica-venti* populations. *Dissertation, University of Hohenheim.*

Melander, B., 1995. Impact of drilling date on *Apera spica-venti* L. and *Alopecurus myosuroides* Huds, in winter cereals. *Weed Res. 35, 157-166.*

Metcalf, E. L., Cress, C. E., Olein, C. R., Everson E. H., 1970. Relationship between crown moisture content and killing temperature for three wheat and three barley cultivars. *Crop Sci. 10, 362-365.*

Mortensen, D.A., Johnson, G.A., Wyse, D.Y., Martin, A.R. 1995. Managing spatially variable weed populations. *In: Site specific crop management. Am. Soc. Agron. 397-415.*

Mortimer, A.M., Ulf-Hansen, P.F., Putwain, P.D., 1992. Modelling herbicide resistance - A study of ecological fitness. *In: Denholm, I., Devonshire, A.L., Hollomon, D.W., (ed.) Elsevier, London. Achievements and developments in combating pesticide resistance, 148-64.*

Moss, S.R., 1985. The effect of drilling date, pre-drilling cultivations and herbicides on *Alopecurus myosuroides* (black-grass) populations in winter cereals. *Asp. Appl. Biol. 9, 31-39.*

Moss, S.R. 1987a. Influence of tillage, straw disposal system and seed return on the population dynamics of *Alopecurus myosuroides* Huds. in winter wheat. *Weed Res., 27, 313–320.*

Moss, S.R., 1987b. Competition between *Alopecurus myosuroides* and winter wheat. *In Proceedings of the 1987 British Crop Protection Conference-Weeds, Brighton, UK, 16–19 November. 367–374.*

Moss, S.R., 1990. The seed cycle of *Alopecurus myosuroides* in winter wheat cereals: A quantitve analysis. *In Proceedings of the EWRS Symposium, Integrated Weed Management in Cereals, Helsinki, Finland, 4–6 June. 27–35.*

Moss, S.R., Clarke, J.H., Blair, A.M., et al. 1999. The occurrence of herbicide-resistant grass-weeds in the United Kingdom and a new system for designating resistance in screening assays. *In Brighton crop protection conference weeds 3, 179-184.*

Moss, S.R., Perryman, S.A., Tatnell L.V., 2007. Managing Herbicide-Resistant Blackgrass (*Alopecurus myosuroides*): Theory and Practice. *Weed Tech. 21, 300-309.*

Moss, S.R., Tatnell, L.V., Hull, R., et al., 2010. Integrated management of herbicide resistance. *AHDB/HGCA Project Report No. 466. 112. Available at: https://cereals.ahdb.org.uk/media/701913/pr466.pdf (Accessed 2019 /09 / 17).*

Moss, S.R., 2017. Black-grass (*Alopecurus myosuroides*): Why has this weed become such a problem in western Europe and what are the solutions? *Outlooks Pest Manag. 28 (5), 207–212.*

Naylor, R.E.L., 1972a. *Alopecurus myosuroides* Huds. (A. Agrestis L.). *J. Ecol. 60 (2), 611-622.*

Naylor, R.E.L., 1972b. Aspects of the population dynamics of the weed *Alopecurus myosuroides* Huds. in winter cereal crops. *J. Appl. Ecol. 9, 127–139.*

Neve, P., Powles, S., 2005. Recurrent selection with reduced herbicide rates results in the rapid evolution of herbicide resistance in *Lolium rigidum*. *Theor. Appl. Genet. 110, 1154–1166.*

OECD, 2018. OECD-FAO Agricultural Outlook 2014-2023. *Food and Agriculture Organization of the United Nations. OECD Publishing: Washington, DC, USA, 2018; ISBN 978-92642-1174-2.*

Oerke, E.C., Dehne, H., 2004. Safeguarding production—losses in major crops and the role of crop protection. *Crop Prot. 23, 275-285.*

Oerke, E.C., 2006. Crop losses to pests. *J. Agri. Sci. 144, 31-43.*

Petersen, J., Wagner, J., 2009. Sneakily increase. *DLG-printing 1, 40-42.*

Piepho, H.P., 2004. An algorithm for a letter-based representation of all-pairwise comparisons. *J. Comput. Graph. Stat. 13, 456-466.*

Piepho, H.P., 2012. A SAS macro for generating letter displays of pairwise mean comparisons. *Comm. Biometry Crop Sci. 7, 4-13.*

Pollard, F., Moss, S.R., Cussans, G.W., Froud-Williams, R.J., 1982. The influence of tillage on the weed flora in a succession of winter wheat crops on a clay loam soil and a silt loam soil. *Weed Res. 22, 129-136.*

Power, J.; Follett, R., 1987. Monoculture. *Sci. Am., 256, 78–86.*

Powles, S.B., Preston, C., 1995. Herbicide Cross Resistance and Multiple Resistance in Plants. *Herbic. Resist. Action Committee Monogr. Available at: http://www. hracglobal.com (accessed 31/08/2019).*

Powles, S.B., Yu, Q., 2010. Evolution in action: plants resistant to herbicides. *Annu. Rev. Plant Biol. 61, 317-347.*

Rosenhauer, M., Jaser, B., Felsenstein, F.G., Petersen, J. 2013. Development of target-site resistance (TSR) in *Alopecurus myosuroides* in Germany between 2004 and 2012. *J. Plant Disea. Prot. 120 (4), 179–187.*

Rueda-Ayala, V., Rasmussen, J., Gerhards, R., 2010. Mechanical Weed Control. *Reference to chapter 17. In Oerke et al. (eds.), Precision Crop Protection - the Challenge and Use of Heterogeneity, 279 – 294.*

Ryan, G.F., 1970. Resistance of Common Groundsel to Simazine and Atrazine. *Weed Sci. 18, 614-616.*

Schweiger, 2018. Prices for crop protection 2018. *Available at https://www.schweiger-handel.de (last accessed 15.10.2019).*

Slepetiene, A., Slepetys J., 2005. Status of humus in soil under various long-term tillage systems. *Geoderma 127, 207-215.*

Stetter, J., 1994. Herbicide Inhibiting Branched-Chain Amino Acid Biosynthesis – Recent Developments. *Chemistry of Plant Protection 10, 2-5.*

Sturm, D., Kunz, C., Peteinatos, G., Gerhards, R., 2017. Do cover crop sowing date and fertilization affect field weed suppression? *J. Plant Soil Envir. 63(2), 82-88.*

Swanton, C.J., Weise, S.F., 1991. Integrated Weed Management: The Rationale and Approach. *Weed Sci. 5, 657-663.*

Swanton, C.J.; Murphy, S.D., 1996. Weed Science beyond the Weeds: The Role of Integrated Weed Management (IWM) in Agroecosystem Health. *Weed Sci. 44, 437–445.*

Thurston, J.M. 1972. Blackgrass (*Alopecurus myosuroides* Huds.) and its control. *In: Proceedings 11th British Weed Control Conference, Brighton, UK. 977-987.*

Turley, D.B., Bacon, E.T.G., Shepherd, C.E., *et al.* 1996. Straw incorporation – rotational ploughing for grass weed control. *Asp. Appl. Biol. 47, 257-264.*

United Nations, 2019. World Population Prospects 2019. *Available at: https://population.un.org/wpp/ (Accessed 2019/09/02).*

Walker, R.H.; Buchanan, G.A., 1982. Crop manipulation in integrated weed management systems. *Weed Sci. 30, 17–24.*

Werck-Reichhart, D., Hehn, A., Didierjean, L., 2000. Cytochromes P450 for engineering herbicide tolerance. *Trends Plant Sci. 5, 116-123.*

Yu, Q., Powles, S., 2014. Metabolism-based herbicide resistance and cross-resistance in crop weeds: a threat to herbicide sustainability and global crop production. *Plant Physiol. 166, 1106–1118.*

Zeller, A.K., Kaier, Y.I., Gerhards, R., 2019. A long-term study of different crop rotations and herbicide strategies: Effects on Alopecurus myosuroides Huds. abundance and resistance development. *Submitted to Crop Prot.*

Zwerger, P., Ammon, H.U., 2002. Importance of weeds. *Reference to chapter 1.2. In: Second ed Ulmer, Stuttgart. Weeds - Ecology and Control, 11-12.*

Danksagungen

Mein größter Dank, gilt meiner verlobten Frau Dr. Yasmin Kaiser sowie unser beider Familien. Ihr seid mir während meines gesamten Studiums Antrieb und Stütze gewesen, habt mich stets ermutigt und gefördert. Vielen Dank für die schönen gemeinsamen Stunden und die Zuneigung, welche mich auch in schwierigen Zeiten nicht verzagen ließen.

Mein ganz besonderer Dank gilt Herrn Prof. Dr. Roland Gerhards, für Ihr entgegengebrachtes Vertrauen und die Vergabe dieses Themas. Ihre fachlichen Anregungen und Betretung waren eine große Hilfe und gaben dennoch genug Freiräume zum eigenen Wirken.

Weiter möchte ich Herrn Prof. Dr. Jan Petersen, Prof. Dr. Wilhelm Claupein und Frau PD. Dr. Regina Belz danken, für die Bereitschaft meine Arbeit zu begutachten und als Prüfer mitzuwirken.

Ebenfalls gilt mein Dank Frau Dr. Carmen Lübken, und der Bundesanstalt für Landwirtschaft und Ernährung, welche dieses Thema gefördert haben. Des Weiteren möchte ich mich bei den Projektpartnern Dupont de Nemours, FMC Agricultural Solutions, Identxx GmbH und Heinz Walz GmbH für unsere gute Zusammenarbeit bedanken.

Frau Alexandra Heyn und Herrn Hartmut Weeber gilt mein besonderer Dank für deren Unterstützung und unsere gemeinsamen Projekte. Außerdem danke ich meinen Studentinnen und Studenten, für Ihre wertvollen Beiträge im Rahmen Ihrer Abschlussarbeiten, welche ich betreuen durfte. Es war eine Freude mit Euch allen zusammenzuarbeiten.

Gegenüber meinen Kollegen im Fachgebiet der Herbologie und am Institut für Phytomedizin möchte ich meine Dankbarkeit nicht nur für das angenehme Arbeitsklima und die gute fachliche Zusammenarbeit, sondern auch für die Hilfsbereitschaft und die gemeinsamen Aktivitäten zum Ausdruck bringen.

Den Mitarbeitern der Versuchsstationen Ihinger Hof und Heidefeld-Hof bin ich für Ihren Einsatz sehr dankbar. Die Umsetzung der Feldversuche wären ohne deren Mitwirken sicherlich anders verlaufen.

Zu guter Letzt möchte ich meinem Bundesbruder Dr. Wolfgang Siegert danken, welcher weite Teile dieser Arbeit Korrekturgelesen hat sowie meiner lieben Landsmannschaft Württembergia und meinen liebgewonnenen Bundesbrüdern für unsere gemeinsame Zeit. Ihr habt sowohl mein studentisches, als auch außeruniversitäres Leben extraordinär komplettiert. Ich bin froh, dass unser schönes Haus mir stets ein Refugium war, ist und sein wird, und ich auch zukünftig mit Euch zusammen an meine Alma Mater Hohenheim zurückkehren werde.

Curriculum vitae

Personal Data

Name	Alexander Kurt Zeller
Date and Place of Birth	February11[th] 1988, Freiburg im Breisgau, Germany

University Education

Nov. 2016 – to date	Candidate Dr. sc. agr. at the department of Weed Science, Institute of Phytomedicine, University of Hohenheim
Oct. 2014 – Nov. 2016	Studies in Agricultural Science, University of Hohenheim: *Master of Science (M. Sc.)*
April 2011 – Oct. 2014	Studies in Agricultural Science, University of Hohenheim: *Bachelor of Science (B. Sc.)*
Oct. 2009 – Oct. 2011	Studies in Economics Science, University of Hohenheim: *canceled*

School Education

2006 – 2008	Angell Gymnasium, Freiburg im Breisgau *Abitur*
1998 – 2006	Friedrichs – Gymnasium, Freiburg im Breisgau
1994 – 1998	Primary School, Freiburg im Breisgau

Alexander Kurt Zeller
zu Hohenheim – Stuttgart, den 25.11.2019

Eidesstattliche Erklärung

Eidesstattliche Versicherung über die eigenständig erbrachte Leistung gemäß § 18 Absatz 3 Satz 5 der Promotionsordnung der Universität Hohenheim für die Fakultäten Agrar-, Natur- sowie Wirtschafts- und Sozialwissenschaften

1.Bei der eingereichten Dissertation zum Thema:

Integrated weed management strategies to control herbicide resistant *Alopecurus myosuroides* Huds.

handelt es sich um meine eigenständig erbrachte Leistung.

2. Ich habe nur die angegebenen Quellen und Hilfsmittel benutzt und mich keiner unzulässigen Hilfe Dritter bedient. Insbesondere habe ich wörtlich oder sinngemäß aus anderen Werken übernommene Inhalte als solche kenntlich gemacht.

3. Ich habe nicht die Hilfe einer kommerziellen Promotionsvermittlung oder -beratung in Anspruch genommen.

4. Die Bedeutung der eidesstattlichen Versicherung und der strafrechtlichen Folgen einer unrichtigen oder unvollständigen eidesstattlichen Versicherung sind mir bekannt.

Die Richtigkeit der vorstehenden Erklärung bestätige ich. Ich versichere an Eides Statt, dass ich nach bestem Wissen die reine Wahrheit erklärt und nichts verschwiegen habe.

Alexander Kurt Zeller
zu Hohenheim – Stuttgart, den 25.11.2019

www.ingramcontent.com/pod-product-compliance
Ingram Content Group UK Ltd.
Pitfield, Milton Keynes, MK11 3LW, UK
UKHW021959190726
13853UKWH00004B/1631

9 783736 972780